John Bruce, O.B.E.
Journeyman Plumber

Journeyman
Plumber
1
Metropolitan
Toronto
1968

By Donald R. Montgomery
Edited by Margaret Dowling

© Copyright 2000.

All rights reserved. No part of this publication may be reproduced, transmitted in any form or by any means, electronically, mechanical or photocopying, recording or otherwise or stored in a retrieval system without the prior written consent of the author, Donald R. Montgomery or Trustees of the Montgomery Family Trust.

Property of the Montgomery Family Trust

Canadian Cataloguing in Publication Data

Montgomery, Donald R., 1920-
John Bruce OBE

ISBN 1-55212-457-6

1. Bruce, John, 1876-1970. 2. Plumbers--Labor unions--Canada--Biography.
3. Labor leaders--Canada--Biography.
I. Dowling, Margaret, 1925- II. Title.
HD6525.B78M66 2000 331.88'1961'092 C00-911181-6

TRAFFORD

This book was published *on-demand* in cooperation with Trafford Publishing.
On-demand publishing is a unique process and service of making a book available for retail sale to the public taking advantage of on-demand manufacturing and Internet marketing. **On-demand publishing** includes promotions, retail sales, manufacturing, order fulfilment, accounting and collecting royalties on behalf of the author.

Suite 6E, 2333 Government St., Victoria, B.C. V8T 4P4, CANADA
Phone 250-383-6864 Toll-free 1-888-232-4444 (Canada & US)
Fax 250-383-6804 E-mail sales@trafford.com
Web site www.trafford.com TRAFFORD PUBLISHING IS A DIVISION OF TRAFFORD HOLDINGS LTD.
Trafford Catalogue #00-0122 www.trafford.com/robots/00-0122.html

10 9 8 7 6 5

John Bruce, O.B.E.

List of chapters

NOTE

All quotations of documents and persons are as they appeared or were recorded. No correction in spelling or grammar has been made.

The photos in this book are taken from periodicals and photographs. Some magazines were first issued in the early 1880's which do not reproduce well. The oldest is the photograph of John Bruce and his mother taken in 1882.

The Author and Editor.

TITLE
John Bruce, O.B.E. Journeyman Plumber

Until the end of World War one, building tradesmen moved easily from city to city, both in Canada and the United States and from country to country. Workers moved to where the jobs were. This practice began in England several centuries ago. A tradesman who could find no work in his town journeyed to where he could secure employment. Indentured apprentices were not permitted to move. Only those who had satisfactorily completed their apprenticeship were free to travel. Hence, the term 'journeyman' also meant that the person was a qualified tradesman.

Some guilds or unions assisted the journeying worker by subsidizing his trip. As he traveled he found accommodation with a fellow guild or union member. The guest signed or marked a 'chit' as evidence that he had been housed at that member's home. The 'chit' would be honoured and a fee of a penny or two would be paid for the lodging by the guild or union.

Thus a tradesman may have worked in several different towns or cities. In North America he could have also worked in both Canada and the United States. Few journeymen in any trade had worked on three different continents as John Bruce had. No one was more entitled to be called a journeyman than he.

On King George the Sixth's Honours List of July 1st, 1946, John was made an Officer of the British Empire. He could have the initials, O.B.E. listed after his name. He rarely if ever did. This honour was afforded few labour leaders. Of course, few labour leaders had served their union for fifty-two years as a General Organizer as John Bruce had. After serving in this capacity for more than a half century

John retired in 1963, at age eighty-seven.

Thus the title of this book‘John Bruce, O.B.E.-Journeyman Plumber’.

For his lifetime of service to the Canadian Government and his war effort King George the Sixth awarded John W. Bruce the title of “Officer of the British Empire”

In 1968, the City of Toronto honoured John Bruce by issuing him Journeyman Plumbers license Number One. John had not worked at the trade since August, 1910

Acknowledgments

Many and special thanks to Carol Rose whose encouragement, help and understanding made this book possible.

And thanks to Charmaine Sayer whose computer skills are acknowledged and appreciated and to

Marvin J. Boede, past President of the United Ass'n of Journeyman and Apprentices of the Plumbing and Pipe Fitting Industry for his help.

Chapter One
The Land of His Ancestors

Dundee, Scotland lies eight miles inland from the sea and is one of its largest cities. Here, in the 1830's and 1840's lived many of the clansmen and their families. Those unfortunates were driven from the land by the self-serving, ersatz Scottish nobility. These were insensitive, greedy vassals of the German-born English King George the Second. In 1745 these 'noble souls' gained the King's favour by joining him in a war of attrition and genocide against their fellow Scots. They became the new English-created Scottish aristocracy. They cared not and had no sense of responsibility for the clansmen. Scots, who for hundreds of years had lived on the land, that were given their new overlords by the English king.

More than a half century later, when raising sheep became a profitable enterprise, these ersatz Scottish nobles drove the clansmen off the land to make space for the sheep. The English, with fixed bayonets, aided in 'The Clearances'. These displaced Scots drifted to the cities and towns to live in squalor in the alleyways, ditches and slums. One Scot, living in abject poverty was William Bruce, an unemployed stone cutter. He had difficulty accepting that living in degradation and destitution was God's will. He questioned the clergy's statements that it was so. The ministers of the established church preached subservience. If the working class accepted the yoke, they would be rewarded in heaven. This promise came with no written guarantee. The rich to whom the clergymen looked for funding approved this gospel.

William Bruce was a questioning man; a thinking man; therefore dangerous and suspect. He could not believe that God had ordained that he and his family must live in an unsanitary and unhealthy hovel and to go to bed hungry each night. He expressed his doubts to associates and neighbours. In such a group there are always one or more gossips, informers who toadied to those who employ them. William Bruce was soon marked as a malcontent and potential trouble-maker.

When he joined the Workmen's Association he was 'black-listed'. Newspapers of the eighteen thirties and forties referred to trade unions as the 'Great Power of Darkness'. When William Bruce joined the Workmen's Association 'polite society' deemed he entered the nether world. He was thus denied what little employment he might have had. With no welfare, or unemployment insurance his poverty worsened. As a malcontent, even what private charity he may have received was denied him.

William Bruce could not understand a society in which a few rich enjoyed all the wealth and privilege while the working class suffered degradation and poverty. The lower classes were deemed to have no rights. The gentry treated their farm animals better than they did the workers and their families. Those with money were insensitive to the misery of the poor around them. They preached that it was God's will. What good churchgoer would question God's will? The clergy echoed these views; views which William Bruce could not accept.

His position was hopeless. When an agent visiting Dundee offered escape, he found a willing listener in William Bruce. Stonecutters and other building trades men were needed in Australia. He agreed to become 'a contract immigrant'. His potential employer would pay for his and his family's passage. Once employed, William Bruce would pay for the cost of moving to Australia. This arrangement was made unlawful in 1902.

Thus, William Bruce, his sons, David and John and two daughters journeyed to their point of departure. While there, the two daughters were 'Shanghaied'. David Bruce told his son this story. He used the word in fashion"Shanghaied" rather than "Kidnapped". The Bruce men were unsuccessful in their attempts to find the girls. The captain would not delay the ship's departure, so the Bruce males boarded, leaving the girls behind. They began their long sea journey to Australia. Although grandson John Bruce said his two aunts were 'Shanghaied', it was not so. He admitted as an after thought, that the two young women did not want to leave Dundee. They had their reasons. One reason was a Scottish lad, named Ogilvy. Later he married one of the sisters and they moved to France. There, she gave birth to several Ogilvys who settled in Paris, Rouen and Nice. It is reported that the other daughter died fairly young.

The sailing ships if weather conditions were favourable could reach Australia in six, seven or eight weeks. The vessel's holds, which on its previous journey may have held wool, lumber, molasses or whatever would produce a profit, were hastily converted to carry passengers. Crowded into this below the deck dungeon, the passengers lived in stench and filth for several weeks. Each traveler was provided with a bunk three by six feet. The bunks were like shelves two or three feet apart. Scattered among these bunks were barrels, boxes and other luggage. There was little passageway. If the weather was bad and the hatches remained closed the stench, smell and stale air contributed to disease and death. The dead were buried at sea. Their departure helped relieve the congestion.

All the immigrants coming to Australia before and at the time of the Bruce family's arrival suffered the perils and dangers of the long sea journey. It took eight to a dozen weeks to travel from the British Isles to Australia. As many as five hundred persons were crowded below deck in four hundred ton, three-masted schooners like the Wellington. Each passenger was given approximately two square feet of deck area. Little wonder, the bunks were four and five high.

The ship sailed along the south Australian coast into Port Phillip's Bay, an area of six hundred square miles. Continuing north to Hobson's Bay, the vessel carrying the three Bruce males, tied up at one of the several piers in Port Melbourne. This dockyard of the City of Melbourne was two miles, as the crow's flies, south of Victoria's capitol. If one traveled by the Yarra River that wound its way through to the city, it was a seven-mile journey. This safe harbour that lay almost forty miles in from the sea provided safe berth for some half hundred vessels.

Tied up at Port Melbourne's many piers were one, two, three and four masted 'tall ships'. Leaving the ship at its final destination, was greeted by all with such joy and sense of relief.

The practice was for a representative of the immigrant 's sponsor to meet the new arrival at the dock, call out his name, and check it against a list. Collecting his charges, he took them to their assigned lodging, usually a barrack type boarding house, owned by the sponsor or a colleague.

The journey led the recent arrivals through a street lined with saloons and brothels that served the crews of the ships waiting unloaded or loaded at dockside. Next they entered an area of shops and small factories. These served the needs of 'the tall ships' and their crews. Had they traveled beyond their immediate destination, they would have stepped out into the village square and the shops and stores that served, for the most part, the local residents.

Later, William Bruce was amazed when he discovered the housing provided for the residents of Port Melbourne. He found them in great contrast to workers' housing in Dundee. There the poor lived in squalid overcrowded row housing. The rooms were low ceilinged, small, with damp walls, and sagging roofs with loose tiles that let in the rain. These unsanitary dwellings lined narrow laneways, with the drain no more than open sewers. The sun's rays rarely heated the hovels. The narrowness of the lanes and the weather made it nearly impossible for such housing ever to lose its damp, unhealthy state.

When William first walked through the residential section of the port village he had difficulty believing what he saw. Land was plentiful in the Colony of Victoria. The immigrant population was determined to use the open spaces to their advantage. The cottages, though small, had front porches, a rear and front entrance as well as their own gardens. He was later to discover slums in Melbourne. They still were a welcome improvement on the hovel in which he lived in Dundee.

The sponsoring agent reminded the 'contract immigrant' of the conditions of his or her agreement. They were to work for the person or firm who had arranged and paid for their fares to Australia. They were provided with work clothes, as many were dressed in rags. Patched clothing that had been worn for seven or eight weeks in confined space aboard ship, was now in tatters. This cost was added to his or her debt. Room and board was deducted from his or her earnings. These were standard conditions of an 'immigrant contract'. No doubt, these were the same conditions imposed upon the Bruce men by their sponsor.

Soon after William Bruce landed in Australia he went to work for his sponsor as a stone cutter. Building tradesmen were in great demand. It was an ideal time to form a union. This they did. To protect their rates, the unionists had insisted that all stone cutters be paid the union rate. Thus William Bruce was able to pay off his debt more rapidly than many other 'immigrant contract' workers were. Immigrant workers employed on jobs where there was no union or the union was too weak to enforce the wage scale were not as fortunate.

As William Bruce had been a union member in Scotland he joined the Melbourne chapter of the Stone Cutters union and became one of its most active members.

David, our hero's father, arrived in Australia as a youth with no skill or trade. He became a shoemaker's apprentice. Shoemakers were not in great demand or considered highly skilled. His failure to find a well-paying job made paying off his indebtedness a more lengthy process. This made his life in Australia more difficult as it kept him indentured for a long period. Still, he enjoyed a much better life than he had when he lived in Dundee.

A year or so later a family friend located the two young ladies. He wrote to William Bruce in Port Melbourne and asked if he would send money to pay their passage to Australia. He was convinced that they had not been shanghaied. He knew that they had not wanted to leave Scotland when they had the opportunity. He had had enough trouble with these deceitful young women. Their father refused to pay their fare to join him.

FOOT NOTE. *William Bruce, John Bruce's grandfather married a woman from Perth, Scotland. John never gave her Christian or surname— nor did he ever mention her again. John did not say she was in Dundee with the Bruce family or that she travelled with her husband to Australia. Whatever happened to her John did not say.*

Chapter Two - The New Homeland

No matter how accurate the descriptions may have been, no one could have prepared William Bruce for what he discovered in Australia. Once landed in Port Melbourne, he studied the neat wooden cottages with their tidy yards; some with vegetable gardens. Flowers bloomed in small beds around the front doors of others. He saw no slums, dilapidated, rat- infested row housing in Port Melbourne. He remembered the squalid, crowded, disease-ridden damp, dark airless hovels with water falling through the ceiling that he and his kind inhabited in Dundee.

He was impressed with the wide roadways and the generous use of space; so much greenery, so many trees and more warmth and more sun here than was in the dreary land of his birth. He was glad he had journeyed these thousands of miles to a land where a man had hope; could be a free man and enjoy a good life. This was the new land, where he would spend the rest of his life and it pleased him to be in the new land.

The Colony of Victoria was as near to being a classless society as possible. One legislator suggested that the name Victoria Colony should be changed to 'Con Victoria' Colony. It wasn't. Still, it had been the notorious prison colony of Botany Bay. William met several emancipators, convicts who had served their time and were now free. For many, the stigma of being a convict still remained, even years later. A few were only guilty of forming a union;or some minor offense. Some men and women had committed no crime at all. Now such offenses would result in probation or a suspended sentence. Former prisoners, their guards, immigrants escaping poverty, persecution and oppression, 'immigrant workers' and those escaping religious persecuation like the Irish were now melded together and their purpose was to form a new society. All remembered too well, the cruelty and injustice of the greedy, self-serving that ruled Britain. They were determined that another Britain would not be built here, 'down under'. The vast majority were united in their resolve that they and their descendants would not be ruled and oppressed by the privileged nobility, the idle rich and ignorant gentry. Every man and woman must enjoy equality and freedom in this new land. After 1872, religion was no longer part of the school curriculum, a natural reaction to Anglican Church monopoly. Religious perception as exuberant in Ireland and elsewhere would not be tolerated here. In this new land, Catholic, Presbyterian, non-conformists, all could own land, hold office; rights denied those of these faiths in the land of their birth. William Smith O'Brien, a Protestant landowner and John Mitchell, a writer for the 'The Nation' were but two of those who emigrated from Ireland to escape religious persecution.

It was a land of informality as only a classless society could be. No need to doff your cap to one who was your oppressor in their new homeland. Grandfather Bruce quickly learned that down there he was no longer William. 'Down under', almost every Bruce was called 'Scotty'. Not that he was any more a Scot than a McBride or McPherson. Nor could anyone explain why it was so. Nor could one learn why every Wilson was called 'Tug'; every Rhodes was addressed at 'Dusty'. Ask why with the surname, a Webb was called 'Spider' and you got no explanation. Ask, 'why' men were tagged with these nick names and you were told, 'that's the way it is.' Or maybe, 'what else would you call a Webb?' If there were three men in the group whose names were Reginald one would be told, ' Well, he's 'Reggie' and that one is 'Reg' and the third was 'Big Reg'. If there were three or more men whose surname began with an 'Mc' or 'Mac', one became 'Old 'Mac', another Young 'Mac' or just 'Mac'. There were 'Red Macs", 'Little Macs', "Highland Macs', and so on. Most men who were part of a group were given a new name. It was a sort of baptism, an initiation; to give the new arrival recognition as a different person now that he was in the new land, a land where the inhabitants wanted to build a new life, a life of equals, where no man was oppressed or victimized by others more powerful or affluent than he, or by society itself.

In the British Isles, the boss was 'Mr. Sykes, Sir'. There in Australia, many seeking to establish an egalitarian society, were addressed as "Hey, Big Red" or "That's been done, Spider'.

Each man had his mate, a fellow more than a friend or brother. For in this man's world, because of earlier times when they were prisoners or indentured servants or bond servants, a man needed a mate to rely on, to be supportive, to share his hopes and fears. The reasons for picking a mate were many and varied. If asked, it is doubtful, that either mate could explain the strong bond between them.

In a crew of workers, such as the gang of stone cutters William Bruce joined, there was a collective pride. If one could lift and carry a larger stone than his fellows, his crewmembers would take pride in his accomplishment He became a minor celebrity, a person his crew could brag about. One of his crew may challenge another crew of stone cutters to a contest; their strongest against his crew's muscle man. If the challenge were taken up, each crew would wager on its champion. Aussies were, and are, known as willing to wager on almost any contest. They love horse racing, betting on their favourite 'hay burner' and on most other sport contests.

If one of their numbers were particularly adept in quickly locating the best seam to split a large rock, his fellows would take pride in his special skill. Such a stone cutter could find the seam within seconds, place his wedge or chisel on the spot tap it with his hammer until the rock split along the seam just as he expected. The cut looked as though a very sharp knife did it. Little or no extra chiseling was needed to square or level the rock's surface. A man with this special talent might be required by his mates to compete, to pit his skill against a fellow stone cutter from another gang. Men took pride in their work and the quality of their crew.

Unlike in Dundee, William Bruce's work place was a friendly one. Much chatter and interchange of ribald remarks and case histories. If one felt like it, he sang. One of their numbers was prone to sing hymns and quoted from his Bible. That is how he earned the nickname,' Deacon'. In Scotland, a building tradesman worked as long as there was light to work by. In Australia, the hours were set out and were less than in the British Isles. The Aussie workers were as productive as those in Dundee or elsewhere in Britain were. Probably they were more so as they took pride in their work and enjoyed a much better work environment

In "Australian Life in Town and Country" published in the early 1900's. E. C. Buley said, ' —immigrants who —contrary to popular belief — are very welcome in Australia, could not try their fortune in a land of better promise. Their material prospects are at once improved, and what is better, they can become masters of their own future. No man is hindered from advancement in Australia by barriers of class distinctions. — The Australian workman fully appreciates these possibilities and the absence of class distinction which they imply, and shows appreciation by an independence of conduct which is very noticeable. It cannot justly be said that this independence is allied to any discourtesy of bearing, but he knows his own value, and is also fully aware of the importance of the political power he wields."

Although this was written half a century after William, "Scotty' Bruce arrived in Australia, these conclusions were being reached daily since the elder Bruce stepped off the ship in Port Melbourne. Sir Gilbert Parker in his book 'Around the Compass in Australia' outlined the process of becoming an 'Aussie'. '"Whether the rural resident is an Englishman or Australian by birth, by the time he has lived in the country ten years, he becomes Australian, as distinct from his former nationality. The Irish man gains direction and industrial confidence; the Scotsman, adaptability and warmth; the Englishman leaves off his insularity and puts on elasticity; and the Australian is slowly evolved." Apparently, William Bruce and if not him, his son and grandson gained such adaptability and warmth.

Bernhard R Wise said; 'To understand Australia, it is necessary to remember that it is the country of the man who works, and not of the man of leisure' [1908]. If there was a country devoid of class distinction, no doubt, Australia had come closer to it than any other.' This was the society that William Bruce had dreamed of. When he settled in Port Melbourne, he thought he had found it. If not his ideal in its entirety, it was close enough to more than satisfy him. He and thousands of Britain's oppressed, down trodden had truly found their almost promised land.

Two immigrants, both characters in the novel, "Voss" by Patrick White, provide two very divergent views of life in the Australian colonies in the mid 1800's. One, Topp, a Professor of Music and a lover of the fine arts expressed his view of these colonists: "I came here through idealism and a mistaken belief I could bring nicety to barbarian minds. Here, even the gentry, or what passes for it, has eaten itself into a stupor of mutton."

The other, Harry Roberts, a poorly educated young labourer, born into a family living in the poverty and degradation of England, held this very different, opinion:" I see nothing wrong with this country, nor with havin' your belly full. Mine has been full since the day I landed, an I am glad."

Others likely held opinions somewhere in between these much different viewpoints.

This was the Australia that John Bruce was born into. For twenty- six years he lived in this egalitarian, semi-socialist environment. It shaped his home life and together with the wide expanse of land, it shaped the boy and the man.

John Bruce at age six and his mother, Amy Bruce {Taylor}. John was her first born.

Chapter Three - School Days

When she was eighteen, Amy Taylor, a native born Australian, married David Bruce, a former 'contract immigrant'. He was eleven years her senior. Until the age of twenty-one he was not in a financial position to marry. Amy with her stocky figure and dominating personality knew she would have few offers of marriage. When David Bruce, a likable, even-tempered man with 'a fair schooling' proposed to Amy Taylor, she accepted his offer without hesitation. They were married either in late 1874 or early 1875.

On February 12th, 1876, their first child was born ,a boy. As the first born, this infant was special in his mother's eyes. They named him John after her father and David's brother. For the next year, he was her only child. Amy, a broad-shouldered and wide-hipped woman, was well suited for child bearing and proceeded to give birth to a child as often as nature permitted. She gave birth to eleven children.

She was a matriarch. Like all her kind, she ruled the family with a firm hand, one that was not always gentle. David, her husband, accepted this arrangement. Relieved of the responsibility of being head of the house, he could devote more time to his union, the Boot and Shoe Workers and his politics. John said, "My first recollection of my father's activities was political. He was active in the struggle to secure some type of representation in parliament." For a time he had been president of the boot workers' union.

Amy and David always needed to make a shilling do the work of two. Amy spent her life constantly cooking, nursing, guiding, mothering and sewing. After her marriage, Amy seldom left the house. John said that he could not remember her leaving the house. One of the rare occasions was when; still protesting she attended John's wedding. Having mothered such a large family, she had little time for a social life. The Bruces lived in a three-bedroom house. Like most houses, it had no basement. The parents shared one bedroom, the girls another, and the boys, the third. With eleven children, the bedroom walls bulged as the two filled with children. struggled to hold their many occupants

In their dress, the immigrants ignored Australia's more temperate climate which made lighter clothing more suitable. Their British heritage, habit and style made them feel the need to wear garments made of heavy serges, wools and tweeds; cloth that wore well and could be recycled time and time again. His mother made Young John's clothes from recycled cloth. Friends and neighbours exchanged clothes for wear and recycling. Pants, shirts, coats outgrown or discarded, were made into clothes for other smaller people.

Like most young ladies, Amy could sew and did so with more skill than most. John was not alone in wearing 'hand-me-down' clothes.

With few exceptions, like other working class families, John and his siblings wore cut-down clothes. Wearing 'hand-me-down clothes' had no stigma. John's only store- bought clothes were his Sunday, church-going suit. When he outgrew this suit it was passed along to his younger brother. John wore short pants until he was ten: then he graduated to long trousers except for play. There were a number of wealthy families in Melbourne, merchants, professionals and businessmen but few lived in Port Melbourne.

They were a small minority and their dress habits were those of the 'Palmy Bastards' {the Aussie term of endearment for English 'would be' gentry} and thus frowned upon by the general population. In the late 1800s the Aussies also called the English immigrants who settled in the 'Outback", 'Jackaroos'.

As an adult, John always dressed fashionably and well. Possibly he was haunted by the memory of the 'hand-me-down' clothes of his boyhood days.

As a preschool boy, John was an active participant in the physical contests that little boys in his neighbourhood engaged in; feats of strength, of endurance, racing and game playing. Physical competition established the 'pecking order'; a. ritual that many species of animals use to determine the social structure, man included.

John was keen to run, jump, fish, and paint, play football and perform feats of strength in which he excelled. His larger size, his quickness, his co-ordination, imagination and 'smarts' soon established him as the group leader, a position he occupied among his fellows all his life.

His comrades looked to John to invent and organize games; sports and such activities as boys in his age group liked to engage in. He seldom failed to do what his mates expected of him. His imagination was used to invent 'make-believe' games. He and his playmates pretended to be characters from some book that had been read to them. Later, they were from books John himself had read. When a little older, he became a better than average swimmer, runner, shot with a rifle, singer, reader and dancer. He would go hunting and fishing with a friend or two. Game was plentiful in those times and what they shot or caught was a welcome addition to the family menu.

John Bruce stood out among his peers. Observant professionals and two grammar school teachers were among the first to notice this. Even as a boy, he displayed special talents and potential. This uniqueness afforded him many opportunities to enhance his status in life. He made good use of them. At an early age he began acquiring confidence in himself. This self-assurance and his quick mind, his eagerness to learn and his willingness to place himself at risk served him well.

In 1882, at age six, John started attending Knox Street Grammar School. It was a two-story brick building with thirty classrooms. A few years earlier, the Government of the Colony of Victoria passed legislation to make school free and compulsory for all children between age six and fourteen. After 1873 religion was not part of the curriculum . At this time the secular school system in the Colony of Victoria was said to be one of the best in the English-speaking world.

Two teachers at the Knox Street Grammar School recognized John's great potential. Miss Young was one and he did not name the other. They saw him as a child with an inquiring mind, an eagerness to learn and a boy easily taught. Miss Young encouraged him to develop his talents, his singing, reading ability for his own pleasure and to add to his knowledge. She urged him to join the choir. He became a member of the Grant Street Presbyterian Church choir and remained a member until 1902 when he left for South Africa.

John liked the discipline of arithmetic. He enjoyed the orderliness of multiplication; especially the 'nine' times table. Every number multiplied by nine, added up to nine, except when a double-digit number multiplied it with 'one' as a second digit. Then the answer may add up to eighteen. 'Nine 'times eleven was ninety-nine which added up to eighteen. As did 'nine' times twenty-one and 'nine' times thirty-one. When you multiplied by fifteen the answer always ended with the last digit being a zero or a five. In later years, his mathematical ability stood him in good stead and singled him out from his colleagues.

When John put his ability to add, subtract, divide and multiply to work, he became keener about mathematics. Figuring out how many feet of fencing was needed to enclose a field, three hundred feet by one hundred with two twelve foot gates and one four foot one was great fun for him. If the field was irregular in size, say, three hundred and twenty -one feet by one hundred and seventy nine with five four foot gates and three sixteen feet ones, it was even more interesting. John was pleased with the challenge if asked to calculate the cost if the price of fence wire was six and half shillings a yard and the total cost was to be shared by two farmers. While in the upper grades, he studied geometry.

John found reading too limited. Reading only simple little stories with words so often repeated and repeated frustrated him. He could say more words than he could spell. He found it disconcerting not to know how to spell all the words in his spoken vocabulary.

Singing and drawing, also on the curriculum, he found most enjoyable as he excelled in both. Yet these activities did not tax his intellect and left his mind only partially occupied. He sang for his own amusement; it was fun, not like schoolwork was supposed to be. His mother kept telling him to study hard. She emphasized that he had to work to learn, he needed to study. John didn't find he had to work hard to excel in his studies.

Grammar too had its discipline. Every sentence had to have a subject and a verb. He soon found that he could create pictures with words. Miss Young encouraged him in the use of adverbs, adjectives, objects and the like. She would often have him read his composition to the class. John found that he liked to be on centre stage. He enjoyed 'showing off' a little. He liked the attention, He was taught by Miss Young to use the word 'very' less and substitute words like 'extremely', 'exceedingly' and 'highly'. John made much use of the dictionary as he sought out new words to add to his ever-growing vocabulary. Miss Young although impressed with his keen interest in words warned him against excesses. She told him he was using words that many people didn't understand. His favourite teacher told him this and he accepted this criticism without comment. He knew that if he wanted his readers and listeners to understand what he was telling them, he must use words they understood. Thus John, while still interested in new words, rarely used them. He remembered Miss Young's caution. Later he wrote and spoke the words his audience understood. He was determined to have his message reach his listeners.

Spelling didn't trouble John. His memory was like a sponge and he filed away what he read or heard. His recall was unusually quick. Even at ninety-three years of age his memory was excellent and his instant recall remarkable. Pronouncing homonyms, words that sound the same but are spelled differently, bothered John. He thought pronunciation should be consistent, like mathematics. There was no order, no discipline. Why were 'pair', 'pare' and 'pear', all pronounced the same and still had different meanings. To John, it made no sense' as did 'their' and 'there'; 'practice' and' practise'. So many words were spelled differently but were pronounced the same. He accepted Miss Young's explanation with regret. Many words in the English language were adopted or adapted from other languages. She told him this and he had to believe it. He had heard the sailors from the foreign ships tied up at Port Melbourne dock talking. He didn't understand what he heard. Yet, once in awhile he thought he heard an English word, one that was not pronounced quite right.

John's interest in word origins and their inconsistent pronunciation made him a good speller. His search for words, pronounced the same but spelled differently improved his ability to spell. Thus, he was an apt contender in any 'spell bee". He often did well when competing with those in higher grades. He also learned that with some of the girls in the upper grades, he should not compete. John, as a young boy, began to recognize his limitations. He learned early not to get beyond his depth to avoid having his ego damaged.

All his or her efforts are rewarded when the teacher has a student with a keen, inquiring mind and an eagerness to learn and he or she learns without apparent effort. To aid John in his development, Miss Young found or created special projects and activities for him. She had him help less able students with their lessons. She cautioned him not to do the work for them but teach them how to do it themselves. Thus he saw himself as an assistant teacher and that pleased him and moderated his ego. It not only helped Miss Young, but it kept John interested and busy and so he was taught how to work with others.

John was the teacher's pet. There was no one at the Knox Street Grammar School who doubted it. None of the other boys in his class would say so in his presence. John was taller, stronger than his schoolmates. Also, his tongue could be sharp, cutting and could be sarcastic. He had a reputation for being quick to use his fists; fists that could deliver punishing blows.

When he was nine, he and the other boys this age were given swimming instruction as part of the curriculum. Every Thursday afternoon in summer the students went to the beach and received swimming instructions. As Port Melbourne was on Hobson's Bay, which flowed into the much larger Port Phillips' Bay, almost every boy already knew how to swim, like most boys living around water do at an early age. The training they received was the use of basic strokes, and diving. Those with more ability than the average were taught to compete. John happily found himself selected for his school's swim team. Teams from the various grammar schools around Melbourne would meet at matches and challenge each other's strengths and skills. John, with his strength and size was the fastest member of his group and won many contests. Even after leaving school, for a time, he continued to compete in the cities in the various colonies.

Reading, writing, drawing, grammar, geography began in the lower grades. English and Australian history were taught only in the senior grades. as was manual training for the boys. The girls were taught dressmaking, sewing, cooking and other skills that would make them good housewives. In those times, few married women had jobs, especially not when large families were the norm.

John's interest in geography began before he knew the meaning of the word. He had seen the flags of foreign ships docked at the piers in Port Melbourne. He noted their uniforms differed somewhat from those worn by sailors on British or Australian ships. He heard them talking in a tongue he didn't understand. His mother told him not to go near or talk to those strange men. He thought that was silly as he could not understand a word they said, so how could he talk to them. His curiosity about the countries, from which they came, stimulated his interest in geography. This subject provided him with some answers he had sought for some time. The geography book told him where they came from and how they lived. He tried to imagine what it would be like to live in these far away countries.

Writing was not John's strong suit. He always wrote larger than life, a Knox Street Schoolboy's deliberately crafted awkward script.

John had heard his grandfathers; Bruce and Taylor speak of the harsh, unfair and unjust persecution of the poor and working class in Britain. He had heard them describe the terrible hunger, and the disease-ridden slums that poverty forced them to endure. His father joined the two grandfathers in their praise of life there, 'down under'. Although somewhat restrained, their endorsement of life in the colony was genuine. Young John had no interest in the history of the British who had so badly treated his forefathers. The greedy, unfeeling rich who thrived by exploiting those less fortunate, held no interest for him. They were the exploiters of the land that his grandfathers and father were so glad to leave and never return to. It was most certainly a country governed by greedy, uncaring selfish men with whom he would not want to live and share a country with. He reluctantly learned what he must about English history to maintain his grades.

Australia's history was more recent, in some ways almost current. Some of those who lived part of the history were still alive; ex-convicts or descendants who were also victims of British injustice and greed. This colony was scarcely a century old. Now his elders had the opportunity to work; to live the good life. This was so different from the slums, the hungry, the pangs of poverty and oppression of the British Isles. John memorized all that was written in his history textbook about Australia.

Science, a subject on the curriculum, was given scant attention by the teachers. Miss Young, like her fellow lady instructors, had interests elsewhere. John enjoyed science, what little he was taught. He read all the pages of the textbook. He did so even though it was not required.

John, always adept with his hands, was a star student in his manual training class. This class and gymnastics were also part of the curriculum. John stood out in the gym class where strength and co-ordination were necessary.

When he became a teenager, John began reading to his father as the latter built shoes. The work was done in a small shop at the rear of their residence. The shop floor space measured about one hundred and eighty square feet. On the east side, two small windows barely provided enough light for David to build shoes. The windows were shiny clean as John washed them twice a week. This enabled David, whose workbench stood under the window, to make maximum use of the daylight hours. The workbench was piled high with leather to be stitched and glued together. When daylight gave way to darkness, a coal oil lamp, on a smoke blackened chain hung from the ceiling above. It could be raised or lowered with little effort. John would sit next to his father sharing what light there was, to read. Smoke from the men's briar pipes, the coal oil lamp and the small coal stove had coloured the walls and ceiling of the small room a deep brownish gray. The notices of past political rallies and sports events and old calendars covered the unpainted walls. They too were covered with yellow nicotine smoke stains and soot. The small iron stove was there to warm the room and take away the chill and humidity. It was rarely lit when the small, low ceiling shop was filled with the men's warm bodies and smoke from the men's briar pipes.

David Bruce was an 'out worker'. The shoe factory cut the leather to size and John's father glued and stitched the boot or shoe together. Many a day after school John would pick up cut leather at the boot factory and brings it to his father's shop.David Bruce was paid for each pair he built, according to an agreed-on scale. With an ever-growing family, David worked late into the night. Young John and his mother helped him with some work in his boot shop. In later years, David stopped being an 'out worker'. It was then he began making 'nice, fancy custom shoes for people of that period who could afford them'. It was quite desirable and fashionable to wear 'fancy custom-built shoes'. Even so, he found it a difficult task to clothe and feed his large family

It was a common practice in boot and shoe factories to have paid readers. John had a fair knowledge of English and was a good reader. Little wonder that David Bruce with his inquiring mind, asked his son to read to him as he worked. His son spoke slowly and clearly, pronouncing each word distinctly in a loud, easily heard voice. What he read, he absorbed like a sponge. Even at age ninety-three, his memory was excellent. He stored away names, places, and events that he could recall eighty years later. His father's friends and political cronies would come to the boot-building shop and join John's audience. These men were born too soon to benefit from the Colony of Victoria's compulsory free school system. Most could read but with some difficulty and were hesitant to read aloud. One man, who listened to the young reader, brought a book to be read. Others followed his example bringing with them books of non-fiction and fiction, magazines and newspapers. David enjoyed listening to the works of Robert Louis Stevenson, Francis Bacon, Thomas Henry Huxley, Henry Spencer, Thomas Hardy and Alphonse Daubet, among others. John and has father especially liked books by Earl Edward George Bulwer-Lytton [1803-1773], such as 'The Last Days of Pompeii', 'Devereux', 'Disowned', 'Pilgrims of the Rhine', 'The Coming Race' and 'Night and Morning'. John said in later years, "These books made a heavy impression on me and made me aware of social conditions that were unjust and unfair. I had an evident desire to be a crusader and going out to change it."

Among David and his cronies, books on socialism were popular. "The works of Karl Marx were being heralded as good reading for those desirous of changing the existing social structure and were in demand at that time," John was to say more than half a century later. These were days when Communism was only a political philosophy, yet to be implemented by any government. The Russian revolution was still a few decades away. Karl Marx's teachings were viewed as a political theory, if not a system. John became a disciple of the author of 'Das Capital', a book widely read, which condemned exploitation by the idle rich. As English masters had exploited most working in the shoe making shop, or their parents, they were ready converts to socialism. Marx had an answer for these villains that had oppressed and cheated John's father, grandfathers, and the other men in his audience. Worker control would right the many wrongs that the workers and poor were suffering from.

John's own words as recorded some eighty years later were: "That's where I got my foundation of socialist knowledge. I read and reread Marx several times through." This exposure to the adult world made him better read, better informed and more mature than his contemporaries. Books read were discussed and sometimes heatedly debated. Most had comments to make about everything read, whether a book of fiction, non-fiction, a magazine article or a newspaper clipping. Young John was exposed at an early age to adult ideas, theories and opinions. Here John learned the rudiments of debate, to cleverly question an opinion. Clever questions done with good intent often served a very worthwhile purpose. Modest, carefully crafted dissent seemed to establish a greater unity and respect among the men sitting on boxes, poorly repaired chairs and the long wall bench. Future leaders in the Victoria Labour Party fashioned, refined and polished their political philosophy and style here: George Sangster, Secretary of the Stevedores' Union, George Walters, Joe Morris who later became leader of the Longshore Men's Union, all became Members of Parliament in later years. All the material he read was written for the adult population, none for a boy, just entering his early teens. As David and his mates were workers and socialists who supported the Labour Party, John became a fellow convert, a mate.

By the time he graduated from grammar school and approached his fifteenth year, he talked, thought and acted much older than his age. The discussions in his father's shop prematurely ended his boyhood days. He thought, talked and argued about subjects that his schoolmates did not understand or have little interest in. It was in this cramped, dark, poorly- ventilated shop that he was to find the philosophy, the ideals and the purpose that he would follow all his life. It has been said that at an early age a person's mind is more receptive, absorbs more quickly what it hears, sees or reads. This was so with young John Bruce. He had prematurely entered the adult world. Despite his youth he spoke, thought and talked like a fully-grown, well-read man.

As a graduate of Knox Street Grammar School he was more mature than others, John then entered the work world. He was well read, wiser than his years but had no marketable skill. Well educated for a boy of those times. Nevertheless he was still an unskilled worker. John was now faced with finding a job to help support the Bruce's ever-growing family. His school boy days were over.

German socialist born May 5th, 1818 and died March 14th, 1883. John Bruce maintained that he was a Socialist, not a Communist. He wrote several books the two best known are the 'Manifesto' and 'Das Capital'. He was a graduate of the universities of the Bonn and Berlin. He devoted himself to the study of economic questions

Chapter Four - Indoor Plumbing

In her book, "Growing Pains" Emily Carr describes her life as a seven year old living in Victoria, British Columbia. She colourfully and accurately traced the evolution of indoor plumbing.

"The art of Emily Carr is exactly the same in her writing as in her pictures. It is the art of eliminating all but the essentials - the essentials for her, that is, the elements that contribute to her impression - and then setting these down in the starkest, most compressed form. She had no wish to paint, or to describe in words, the things around her as other people saw them; the camera and the phonograph could do that: it was not work for the artist. What she wanted was to study the things and the events which she felt contained material, until she had extracted that material and thrown everything else away."
(B...K. Sandwell in Saturday Night)

*"**Waterworks***

Those Victorians who did not have a well on their own place bought water by the bucket from the great barrel water cart, which peddled it. Water brought in wooden pipes from Spring Ridge on the northern outskirts of the town was our next modernness. Three wonderful springs watered Victoria, one on Spring Ridge, one in Fairfield and one at Beacon Hill. People carried this sparkling deliciousness in pails from whichever spring was nearest their home.

My father was so afraid of fire that he dug many wells on his land and had also two great cisterns for soft water. Everyone had a rain barrel or two at the corners of his house. The well under our kitchen was deep and had a spring at the bottom. Two pumps stood side by side in our kitchen. One was for well water and one was a cistern pump - water from the former was hard and clear, from the cistern it was brownish and soft.

When Beaver Lake water was piped into Victoria, everyone had taps put in their kitchen and it was a great event. House walls burst into lean-to additions with vent pipes piercing their roofs. These were new bathrooms. With the coming of the water system came sewerage. The wretched little "privies" in every backyard folded their evil wings and flapped away - Victoria had at last outgrown them and was going stylish and modern.

Father built a beautiful bathroom. Two sides of it were of glass. It was built over the verandah and he trained his grapevine round the windows. The perfume of the vine in spring poured through the open windows deliciously. Father had tried to build several bathrooms before Beaver Lake came to town, but none of them had been any good. First he used a small north room and had a cistern put in the attic to fill the bathtub. But hot water had to be lugged upstairs in a bucket and anyway the cistern froze every winter, so that bathroom was a failure He had made us an enormous, movable wooden tub like a baby's bath big enough for a grown-up to lie in flat. It was very heavy and lived on the back verandah. Bong brought it into the kitchen on Saturday nights before he left for town. It had to be filled and emptied over and over by the ladies of the household with a long-handled dipper until all the family had had their baths. Besides this Saturday night monster there were wooden wash tubs painted white which lived under our beds. We pulled these out at night and filled them with cold water. Into this we were supposed to plunge every morning. This was believed to harden us; if your nose were not blue enough at the breakfast table to guarantee that you had plunged there was trouble.

Father later tried a bathroom off the washhouse across the yard. A long tin pipe hung under the chin of the washhouse pump and carried cold water, but hot water had to be dipped out of the wash boiler on the stove. This hot bath arrangement was bad; we got cold crossing the yard afterwards. So the wooden tub was invited into the kitchen again each Saturday night until we became "plumbed".

It was glorious having Beaver Lake pour out of taps in your kitchen and we gloated at being plumbed. Mothers were relieved to see wells filled in, to be rid of the constant anxiety of their children falling in and being well drowned. Everyone was proud and happy about this plumbing until the first hard frost.

Victoria used to have very cold winters. There was always some skating and some sleighing and spells of three or four days at a time when the wind from the north would pierce everything. Mother's milk pans in the dairy froze solid. We chopped ice-cream off the top to eat with our morning porridge Meat froze, bread froze, everything in the house froze although the big hall stove was red hot and there were three or four roaring grate fires as well. Windows were frosted in beautiful patterns all day and our breath smoked.

It was then discovered that plumbers, over-driven by the rush of modern arrangements, had neglected to protect the pipes from frost. Most of the bathrooms were built on the north side of the houses and everything froze except our deep kitchen well. Neighbours rushed to the Carr pump, spilling new snow over Mother's kitchen floor till our house was one great puddle and the kitchen was filled with the icy north wind. Everyone suddenly grumbled at modern plumbing. When the thaw came and all the pipes burst everyone wished Beaver Lake could be piped right back to where it came from.

Once Victoria had started modern off she flew with all sorts of newfangled notions. Cows were no longer allowed to roam the streets nor browse beside open ditches. Covered drains replaced the ditches and, if your cow wandered into the street, she was impounded and you had to pay to get her out. Dogs were taxed but were still allowed to walk in the streets. A pig you might not keep within so many yards of your neighbour's nose. Jim Phillips had to give up his James' Bay farm and remove his piggy to the country. Small farms like his were wanted for cutting into city lots. You never knew when new lumber might be dumped on any piece of land and presently the lumber was a house and someone was moving in."

Indoor plumbing was still a novelty when John began his schooling. The plumbing trade was in its infancy. Water was just being piped indoors. Disposing of liquid waste was a problem for which many sought a solution. Cesspools, ditches filled with the waste created by running water created epidemics, Thousands died from pollution, including Prince Albert, Queen Victoria's husband.

John began his apprenticeship when fear of plagues was very real. Those with knowledge of sanitation were primarily concerned with the disposal of liquid waste, especially with the refuse coming from the indoor flush toilets being installed in the homes of the more wealthy. (See Chapter Five — The Apprentice) Plumbers of that day were merely handy men in the clumsy process of becoming tradesmen. Unfortunately those who were deemed to be qualified plumbers knew little about the handling of liquid waste, so were unable to be helpful to the apprentices with whom they worked. John Bruce with his inquiring mind and his keen desire to learn, took lectures at a vocational school and at the University of Melbourne to learn what his apprenticeship did not teach him.

If there are few men born in the right decade. John Bruce happened to be one such man. When he finished his apprenticeship indoor plumbing was the coming thing. No man was a better qualified apprenticeship candidate. No accredited journeyman had more knowl edge. The next chapter may, to some, seem to provide too much technical detail about the plumbing industry. The author thought this was necessary to acquaint the reader with the rather remarkable technical advances made in the latter part of the nineteenth century. Few made as big a contribution as Thomas Crapper, the inventor of the first successful flush toilet. This invention created many problems that took years to solve.

Few have made such a notable contribution to the Plumbers union as John Bruce. Hopefully, this detail will help the reader appreciate the large part in protecting the health of the nation was made by the world's plumbers. No other building trade can make such a claim.

Chapter Five - The Apprentice

In the late 1870's, there was renewed and growing concern about preventable diseases. Some enlightened doctors, like Dr. Derby of Boston, Dr. Stephen Smith and Baldwin Latham, a sanitary engineer were warning the public about sickness caused by pollution. Water polluted by leaking cesspools, faulty drains. damp, ever-wet basements, outhouses cleaned by 'honey' wagon' drivers with vehicles that leaked and the like, were all potential sources of sickness. Gravity fed water systems with reservoirs in the attic added to the danger. There was more liquid waste with excrement flowing into ill-designed and poorly installed drainage systems. In earlier times many drains were made of wood and had no sealed joints. Liquid waste leaked through the joints, polluting the track along the drain. The homeowner's prime concern was to get the liquid waste away from his house and drain his liquid waste off to some pond, ditch, swamp or ravine. There he trusted in God to turn it into a harmless mush.

Dr. Derby of Boston made many attempts to warn the public of the danger of disease, like typhoid fever, scrofula, uterine diseases, pulmonary diseases, especially the early stages of consumption and such sicknesses caused by pollution. He kept telling the citizens, "Whether the vehicle be drinking water made foul by human excrement, sink drains, or soiled clothing, or air made foul in enclosed places by drains, decaying vegetables or fish or old timber; or, in open places, by pig sties, drained ponds or reservoirs, stagnant water, or accumulations of filth of every sort, — the one thing present in all these circumstances is decomposition. It is in these decomposed substances that diseases are born."

A statement made to create alarm, read as follows: "We roll up your eyes and stand aghast when contemplating the horrors of war; as yet the mortality of war is trifling compared to the mortality of preventable diseases. England, in twenty- two years of war lost 79,700 lives. In one year of cholera, she lost 141,860 lives." *{From the Atlantic Monthly of September, 1875}*

By the 1890's, most municipalities were forced to install sewage and sewage disposal systems. It was a new industry. The public was told and believed that "Everything that can affect the health of the poorest and most distant of our neighbours may affect us, particularly the spread of disease in closely-built towns is more often than not, from the poorest classes upward, so that many a patient falling ill of contagious or infectious disease in the slums of the city, becomes the centre of a wide infection. The health of each is important to all and all must join in securing it."

The earlier and various experimental approaches to waste disposal and excrement yielded, for the most part, unacceptable results. Few succeeded in reducing disease and the mortality rate.

A century earlier, Dr. Benjamin Rush, an eminent physician was so certain that the means of preventing pestilential fevers were, as he said, "as much under the power of human reason and industry as the means of preventing the evils of lightning and common fire". He also said that he looked forward to the time when "the law should punish cities and towns for permitting any sources of malignant and bilious fevers to exist within their jurisdiction." Obviously he was a man ahead of his time, also, a man who believed in lightning rods and fire prevention.

By the 1880's, many people believed the municipality should be held responsible for disease control. There was a demand for sanitary engineers to design safe sewage systems and supervise their installation. Sewers, drains, and waste disposal basins must be built. The plumbers must acquire new skills. Some of the older ones found this difficult. John Bruce, still an apprentice, was eager to learn the new skills, to acquire knowledge the journeyman plumbers themselves were still trying to learn. Some found it difficult to teach their apprentice helpers what they themselves were attempting to understand.

When John graduated from grammar school he had no idea that he was to become a plumber.

The Bruce family, who always scraped to make ends meet, now saw John as a potential wage earner. He had to go to work, as the family needed the money. He applied for work in the boot factory. Its owner, one of those heroic employers Horatio Alger wrote about, told John, "You're too good a boy to work in a factory. You should go into some trade." He arranged with his brother-in-law, a master plumber who owned a plumbing and sheet metal shop, Mr. Marsh, to take John on as a plumber's apprentice. In 1891, John became an indentured plumber's apprentice. Again, John's special qualities were recognized. This was one of the most fortunate happenings in John's long life. In 1891, at age fifteen, John began his career in the plumbing trade, one that was soon to require more training and skill than ever before. This trade was to change drastically in the next dozen years. In 1969 John still had his initial apprenticeship papers with him in Toronto.

Amy Bruce woke a sleepy John up at five-thirty in the dark of the morning. It was time to feed him and get him on his way to Melbourne. She was determined that he be on time for his first day at Marsh and Company.

With his recently acquired but used lunch bucket in hand, John began his three mile walk to his place of employment, the plumbing and sheet metal shop of Marsh and Company. The crispness of the morning air under a cloudy sky made John walk briskly. He covered the distance in fifty minutes. On most workdays, John spent his halfpenny to take the horse car from the Port to the shop in Melbourne.

John was the second person to arrive. Mr. Plunket, the clerk was there already, seated on his high stool before the oak bookkeeper's desk. A subdued, young John introduced himself. The short balding clerk examined him with interest. He took his own sweet time to do so John thought.

"Can you count and add?" Mr. Plunket asked. No 'good morning', 'hello', or 'how are you?' Just this simple curt question.

"I always got good marks in arithmetic," John answered.

"Good. Today you can show me how good you are."

Soon after, Mr. Marsh arrived, dressed in a three-piece tailored suit and carrying a walking stick. He had interviewed John in his brother-in-law's office at the boot factory, so he knew John by sight. "Welcome to Marsh and Company, John. I hope you will like it here."

Before John could reply, the clerk said, "Today I could use this young lad to help me take inventory. He tells me he can count and add, an ability no one else around here seems to have."

John could not tell whether Plunket spoke in hope or desperation.

"Good way to start. Get to know what's important in the shop. Make sure you introduce him to the crew as you go," said John's new and first employer.

For the rest of the morning, John reached into bins to count couplings, elbows, tees and fittings of many kinds. He read aloud the carefully printed words on the signs attached to each bin or pipe rack.

He made his count with care and called out the number. For the first six bins, Plunket would follow him and make his own count.

When satisfied John could count accurately, he just stood, clipboard in hand and wrote the number down that John had given him.

The bell clanged announcing the lunch break. "Come," the clerk said. John followed him back into the office. Plunket heated water for tea then went to the drawer of his desk and brought out his lunch. It was wrapped in a white cloth that looked freshly laundered. John reached down and retrieved the lunch bucket he had left by the door when he entered the office five hours earlier. He had expected to be sitting out on the wood benches at the lunch table in the shop with the apprentices and older men. They were busy eating. Some had a bucket of beer in front of them.

John sat, ate his cheese sandwiches made with his mother's homemade bread, discreetly watching Mr. Plunket. He began that day by addressing the clerk, as 'Mr. Plunket' and continued to do so for all the days he worked with Marsh and Company. The new boy sat, munching his food, sipping his tea, not saying a word. He wondered if the silence would ever end. The clerk ate slowly. He stirred his tea with deliberate thoroughness as though the sugar must be completely dissolved and evenly distributed throughout the whole cup.

Plunket said, without looking up, "All but one of those men and boys sitting at the lunch table out there are from the sheet metal shop. Most of the time, the plumbers, pipe fitters and their apprentices, go directly to the work site. They don't come in here more than once or twice a week. They come in to get the new work orders and the supplies they need for the next job. Soon there will more working in the sheet metal shop. Mr. Marsh has bought small hand shears, rollers, edgers and seamers to make metal containers for the food producers. There is a growing demand for containers. More companies are exporting food and need more containers. Mr. Marsh is intent on getting the costs down and getting more customers."

He took a sip of his tea, now getting cold, before adding, "He plans to make them but not use sheet metal tradesmen to build them. The way he is setting up the work assignments, there won't be much skill needed." He now looked up at John and smiled. "But you're going to be a plumber and there is a need for more plumbers." Then he lapsed back into the silence that had filled the room.

John saw no need to respond.

After lunch, the counting began again. This time, John stood and read the signs on the pipe racks. Some pipe he had to measure for length. He quickly learned which one he must measure, as the clerk would hand him one end of a tape. Then John would march down the pipe rack and call out the pipe's length in feet.

At five o'clock, the bell rang again, marking the end of the workday. John had completed his first nine hours as an indentured apprentice. He had worked as a junior clerk, a white-collar worker, counting, counting.

The day had not been wasted. For John had spent his time trying to memorize all that Marsh and Company kept in its inventory. He planned to spend more time memorizing every item in stock. As he hurried home, he realized that Mr. Plunket knew more about Marsh and Company than anyone else. The clerk was a good man to become associated with. Though not much of a talker he was a pleasant person, one who expected you to work diligently. Working hard came naturally to "Young" John Bruce. His relationship with 'Mr. Plunket' was bound to be a healthy one.

Once Mr. Plunket had introduced him as "Young John" that name stuck. Ever after, he was 'John' to his equals and 'Young John' to his elders.

It was the days of iron pipe and lead drains. Pipe had to be cut and threaded. All this was done in the 1890's, by hand. Strength was needed and the larger the pipe's diameter the more muscle that had to be used., Because of his strength John was given the task of cutting and threading pipe. He took pride in this work and he produced more than the other apprentices given the same assignment.

Water systems were very basic and indoor toilets were in their infancy. Water lines and drains passed as sewers. In much of John's first year as an apprentice he worked with a wiry old Scot journey plumber. Everyone called his mentor 'Sandy'. He was an expert in repairing smelly, leaky drains. When the odour became so foul, unbearable, persistent and intolerable, the homeowner would cry for help. If his plea was made to Marsh and Company, Sandy, if he were available, would be sent post haste.

Quickly locating the source of the smell was vital, not only for the homeowner's benefit but also for the plumber and his apprentice as well. The method for locating leaks was simple and ingenious. The plumber and his apprentice helper would pump smoke into the drainage system. The smoke would escape through the leaks, cracks and loose joints in the drain, even drift out through flag stone walks and elsewhere.

Plumbing supply catalogues advertised 'Smoke Drain Testers'. A popular choice was Barron's 'Air Pump and Smoke Generating Machine'. Oiled cotton when set afire smoldered. The cotton waste that provided the smoke sold for the price of seventeen shillings per hundredweight. (CWT.)

For some jobs Sandy would use the smaller single action tester, such as the 'Wilkinson Drain Grenade' or the 'Kemp Drain Rocket', the latter priced at eight shillings six pence per dozen

Sandy told "Young" John that sometimes none of these smoke devices worked. If the homeowners had waited too long before calling for a plumber there would be too many cracks and leaks, so many places for the smoke to escape, that the work area would fill with smoke and you could not see. Your eyes would water and you had to get out in the open air. When this happened, the crew would switch to peppermint oil. The 'Peppermint Oil' test only worked when the plumber and/or his apprentice had very sensitive noses, and could distinguish the smell of peppermint from all the foul, horrible smells seeping out thorough the many leaks and cracks.

In this foul smelling, damp, sometimes contaminated environment, Sandy and John would work to repair the faulty drains. They sealed cracks, plugged leaks and replaced those drainpipes that were beyond repair. The fault was that many drains were installed by men with little knowledge or understanding of a safe method for getting rid of liquid sewage. The volume of waste continued to increase as more indoor water systems and toilets were installed. The late eighteen hundreds were dangerous days for plumbers and their crews. John found it difficult to believe that the problem of foul smelling, disease-breeding drainage systems could not be solved.

Mr. Marsh and those in his age group knew little about drainage, sewers and liquid waste disposal systems. The journeymen who had trained them had little or no knowledge of sanitation engineering. John Bruce realized that he would have to find out where to take a course on sewage, drainage and liquid waste disposal systems; to find a vocational training school and/or local university that provided such training. When he finished his apprenticeship he did this. He didn't do it before as he thought it would reflect upon Mr. Marsh and the shop's journeymen.

For obvious reasons, John welcomed the move from drain repair to his next level of training. At this stage he was taught how to install water and toilet systems. John studied his textbook and examined and re-examined its pages.

Thomas Crapper, an English plumber and manufacturer, made a most notable contribution to the public's well being. It was what today, with its various improvements, is called the 'flush toilet'. In the late 1800's it was sold under the name, 'T, Crapper - Valveless Water Waste Preventer' It was a great improvement on the Jenning's Closet with its pull up metal plug. This plug was pulled up to flush and pushed down to close off the flow of water. Early in this stage of his training, John was called in to fix a Jenning's Closet toilet. The homeowner told him that the water kept running, not much, but there was a continual leakage between the two tanks. John turned the metal plug around with the hope of improving the fit. That failed to solve the problem. He removed the plug and examined it. That was of no help. Making an excuse that he must go to the shop, he left. He sought out his first mentor, Sandy.

John told this much-experienced plumber of his problem. Sandy snorted, laughed and said that the Jenning's Closet rarely fit properly. A metal on metal fit more often than not permitted water to leak from one chamber to another. Sandy said John could replace it but the odds were it would leak as much as the one it replaced. John returned to the scene of his frustration, and apologetically explained the problem. The Homeowner shook his head and said, "That's what the two other plumbers told me. Why'" he asked, "do they sell toilets that leak?' John had asked Sandy the same question. He quickly gave the answer Sandy had made to his inquiry.

"Sir, it was the best available at that time."

Thomas Crapper of English working-class parents began his apprentice as a plumber at age eleven. No one was more suited for this trade than he. He held more patents on toilet fixtures, drains, traps and the like than anyone, before or since.

After his graduation from the Knox Street Grammar School, John became more active socially. His family, that is Amy, were temperance people, save and except Grandpa William Bruce who, according to his grandson John, 'liked his beer'. Naturally then, John became very active in the temperance movement.

When sixteen years of age, John became an ardent supporter of the International Raccabites, a British institution that had established branches in Australia. He also became an active member of the 'Band of Hope'. This temperance organization aimed for "Total Prohibition'. The former, although a temperance group for young people, was also a fraternal organization that provided some benefits for its members. It organized social evenings, usually on Saturday night; they held sing-songs and sponsored amateur stage productions. John admitted he had 'a good voice and he started to take part in the 'amateur theatricals.' These social activities, John said, "were one of the avenues by which the boys and girls of my period tried to fill in their leisure hours."

They produced plays illustrating the happiness of sobriety and the misery of drunkenness. 'The Age of Reason' was one of their often-produced theatrical presentations. These dramatic productions would last two or two-and-half hours. For some, tickets were sold; for others, they 'passed the hat' in which the audience was asked to drop in money. A poor play would result in a poor collection.

In the 1890's, the temperance movement was independent and not part of any church group. John who always advocated temperance, expressed his faith in these words, "My position on the booze trade was developed quite consciously because I had studied it but could never understand it. Why people drink because it does nothing for them. They seem to go berserk, become irrational, at times objectionable. Having worked around distilleries and breweries in my period of training, I was surprised that anybody would drink the stuff. If they saw the conditions under which it is manufactured they wouldn't. Men working in great heat, naked save for trunks and in their bare feet, their perspiration running off them into the malt." Another time he said, "During this process {joining the temperance movement} I became a great prohibitionist; the drink traffic was responsible for privation, suffering that people were undergoing. I became an ardent supporter of the temperance movement."

At fifteen and a recent graduate of Knox Street School, John liked to sing. The choirmaster of Christ Church Cathedral in Melbourne heard him sing. He was impressed but he noted John had a nasal twang in his voice. It was the result of him breaking his nose in a fall when he was nine years old. The choirmaster knew that a simple surgical procedure could remove the cause of John's nasal twang. He persuaded his church to pay for such an operation on John's nose. They agreed but only, if he attended their school to study music. This appealed to John who saw it as a chance to have a career in music as a singer. His mother, the matriarch, and a staunch Presbyterian, said, "I will not have my son taught Catholicism by some Roman Catholic priest." John's reply was, "But Christ Church Cathedral is High Anglican, Mother." Her tart, quick retort was, "It's the same thing." She flatly refused to reconsider her decision so John lived with his nasal twang all his life. Otherwise he may have become as famous as his Knox Street School fellow student Bill Baker, who as an adult, became a celebrity throughout Australia. John believed that he too could have become a professional singer if he continued in school Of course, he would not be earning any money to help with the Bruce family finances. This factor may have played a part in his mother's negative response.

John also went to dances as frequently as he could.

The stevedores and others struck in support of the sheep shearers in 1893. Approximately a thousand strikers and their supporters assembled near the gas works where John Bruce was working. Among them was his Grandfather Taylor. They adopted what the authorities deemed to be a menacing attitude. They marched through the city. Colonel Tom Price, afterwards known as 'Bloody Tom Price', called out the state militia. He issued each man forty rounds of ball ammunition. The Colonel and his 'Saturday Night' soldiers lined up; blocking the advance of the strikers. The young riflemen in their ill-fitting uniforms were nervous and untrained for such action. Their training had been to march about the parade grounds forming four and the like, a few route marches and some time on the rifle range. Colonel Price poorly trained himself, ordered the civilian marchers to stop. They did not obey his tremulous command. The men kept moving forward. Rather than tactfully giving ground, the panic-stricken Price ordered his men to fire. Knowing little other than to obey, the uncertain militiamen fired. Shocked by what they had done, they made a somewhat disorderly retreat.

Shouts, curses and cries of pain filled the air. When the smoke cleared, there were several strikers and their supporters injured and seven were dead.

The official government reaction was to appoint four boards. The boards numbered four to ten persons each, half employer representatives, and half worker representatives. As a result, a flurry of new labour laws was enacted in the decade that followed this senseless murder of seven strikers and strike supporters.

For John Bruce the result was different. Too young to join the union, he took advantage of what he knew. He joined the Socialist Party.

To advance himself in his new, inexpensive social world, John took elocution lessons. His exposure to this training refined and polished his natural ability as a public speaker. He was an apt student. He acquired or was born with a natural stage presence. He was sensitive to his audience. He could read their reactions. Unlike many speakers who speak to hear themselves talk, John spoke to deliver a message. He related to those who were listening.

He could read them expertly. His stage presence and his loud clear enunciation and skill as an elocutionist served him well. His voice could be heard in the open-air forums where he often spoke.

He would go out on a Sunday afternoon to 'soap box'. He would stand and propagandize on the Yarra River banks of the Melbourne Botanical Gardens. Those speeches of his gained him the reputation. By the age of eighteen I was known as 'The Boy Orator".

As John said years later, "The political movement of that day was taking on young members. I had become quite an effective soap box speaker; and I was used in that capacity."

"I was used as a propaganda speaker, doing plenty of 'soap box' work because we had no halls available to us and we also took advantage of the Saturday evenings. They were known as the shopping evenings when there were people in the streets. We would do all we possibly could to hold a crowd."

He admitted to being a 'dramatic' speaker. He could have been an excellent actor. There is no record how good his stage performances were in the amateur theatrical he was in. He spoke with emotion, with sincerity, spoke as a 'true believer'. At times, to emphasize a point, he would whisper softly. His audience had to strain to hear him and they leaned forward in their seats to hear his every word.

In the third year of his apprenticeship, two of his mates invited him to go with them to Western Australia. It was the least developed colony but occupied a third of the continent of Australia. Minerals had been discovered there, so some saw it as an opportunity to get rich quick by finding some valuable ore deposit. So, it was with Martin and Brown. John was keen to join them. He had a second reason to visit these barren lands with its black, stagnant water. He had developed a water filter. It could turn foul tasting water into a more drinkable liquid at a rate of a gallon an hour."

He spoke with his father who told him not to go. His friend and ally, Grandpa Taylor, was very receptive to the idea. He said, "John, I came to this country to find a new way for myself and I see no reason why you shouldn't go to another country to make a new place for yourself if you think you can better yourself."

When his mother learned of John's contemplated plan to abandon his job as an apprentice and travel to this remote western colony, 'she blew her top'. She realized what John contributed to the family income would be gone with him. Not one to waste time in argument, she devised a plan. A friend of the family was Jake Edwards, a magistrate. With his collaboration, Amy Bruce had a police sergeant arrest John. At that time, John was working at a site opposite the courthouse. The police sergeant went across the street and arrested this would-be traveler to Western Australia. He was locked in the police cells for three hours. Then he was taken into the court and placed in the dock. Magistrate Edwards charged him with planning to break his apprenticeship agreement.

As John was to say, almost seventy years later, "They could pull it off then but not now." Under such duress, young John swore he would honour his apprenticeship agreement. Having made the commitment to honour his obligation he was released. It could be that for all his life he regretted being thus hornswoggled. His mates, Martin and Brown struck it rich and found one of the largest gold deposits in Kalgoorlie, Western Australia.

John said he had a sneaking suspicion that Mr. Marsh, his employer, if not a fellow conspirator, had knowledge of his mother's ploy.

Mr. Marsh persuaded John to join the militia. This was strange, as the Bruce family was pacifist. Nevertheless, John joined, first in the drum and bugle band and he played both instruments. '"I saw the advantage of gaining knowledge that was absent from my trade [training], I transferred to the engineers," He did so as he was keen to learn new skills.

In 1902, he was promoted to the rank of sergeant in the engineers. Although he may have joined the militia to please his employer, John never said so. His explanation was that 'as a member of the militia he would be paid fifty dollars a year which in those times would provide a young man with an adequate wardrobe. "This enabled me to clothe myself." Maybe, John remembered as a schoolboy wearing 'hand-me-down' clothes.

As a member of the Engineers, he was taught how to use the army issue bolt-action rifle, a 303 Enfield. John's strength enabled him to hold the gun steady on target. This and his sharp eyes made him an excellent marksman. His talent was noticed. He became a member of the engineers' rifle team. Members of this militia fired the standard army issue Enfield rifle and used army issue 303 ammunition. No fancy custom carved stocks with elaborate gun sights and small bore rifle for these competitors.

The targets were placed one hundred and two hundred yards away. The individual targets could be raised and lowered from the butts on the target range, so the judges could mark the scores. The smaller the cluster and the nearer it was to the true centre of the target, the higher the score. The shooters would fire from three different positions, standing upright, sitting or lying prone. To be a successful competitor one must be fully aware of the peculiarities of his rifle. He must know if it shot high, low, left or right of dead centre and by how much. Success depended on this knowledge. The reputation of each militia unit lay heavy on the shoulders of its champions.

"My employer was also one of the leading rifle shots in the company of militia engineers and he encouraged me in this activity. Because of this we were able to utilize our Saturday afternoons by shooting"' were the words he used to describe how he spent many of his Saturdays. At noon on the sixth day of the week, at the end of the short four hour workday. Mr. Marsh and John would eat a quick lunch. Then in sun or in a slight drizzle they would proceed to the militia rifle range located near the Melbourne Armouries. Entering the stores, they would take the rifle assigned to each, draw one or two hundred rounds of ammunition and walk to the rifle range.

One would go down to the butts to raise and lower the target. The other would place himself at one hundred or two hundred-yard marker and fire five rounds at the target's centre. They proceeded to the butts where they would examine and record the hits. Then, the other would take his turn. fire the five rounds and walk back to the butts, examine the target and tally the score. Each commented on the other's accuracy or lack it; thus sharing their enjoyment of the sport.

Despite the considerable difference in age, they became mates in this process. Well, mates after working hours for the employer-employee relationship must remain in place. They would do this for two and half to three hours. When finished they would return their Enfield rifles and unused ammunition to the stores. They walked to the horse car line, Mr. Marsh going into Melbourne and John back to Port Melbourne. Sometimes there would be other militia marksmen there and they would become impromptu competitors. Each shooter put up six pence, winner taking all.

In addition to these informal contests, there were more valuable prizes for shooting. "I, like others, got my share of cash prizes that were very beneficial," John commented when he was eighty-nine years old.

John's size and strength enabled him to be a competitive speed swimmer. Entering two and three hundred-yard races, he gained 'considerable prominence'. He competed as an equal with the well-known swimmers of the 1890's, like Cavell and Beaucare. He competed in the back stroke, breast and Australian Crawl events. In the 1890's, swim meets were held outdoors in a sheltered area of some river, pond or lake. No indoor, controlled temperature, smooth pools for these swimmers. The temperature could vary ten degrees or more from one meet to another. In its protected locations there were no great waves but the water could be choppy and cold.

As a member of the rifle team and the local swim team, he traveled to distant cities to compete. He journeyed to meets in New South Wales, Queensland, Southern Australia and New Zealand. These trips gave him ' a general background of the country'. It also gave him 'itchy feet' and the desire to travel. As he put it, "In addition, I got a number of trips to far away places in other parts of Australia in competitions, which gave me an opportunity to see the country and also to meet workers and others in the parts I visited. This stimulated my activity in the trade union movement".

When nineteen years of age, he began dating Alice Ripkey. She was one of three daughters of the local dry goods storeowner. In all, there were six children of this Jewish father and Protestant mother. The Ripkey's welcomed John into their home. They would often gather around the piano and the eldest Ripkey daughter played while the others sang the popular songs of the period. John had by far the best voice and the Ripkeys often persuaded John to sing solos. This was not a difficult feat. John always enjoyed being on centre stage.

When a singing contest was announced in Melbourne, the Ripkey daughters urged John to enter. John played the shy one, saying "You must get an entry form from the contest secretary's office. It closes before I finish work. I can't take time off to go there during office hours."

"No problem there," said Alice. "I'll go and get one. You fill it out, sign it and I will hand it to the Contest Committee." Needing little encouragement, he entered the contest. He selected a new song that was popular at the time, 'The Holy City'. He rehearsed with the piano-playing Ripkey daughter. The night of the contest, the Bruce and Ripkey families were out in full force, except for John's mother, Amy. John entered from the wings, stood centre stage, looking taller than his five feet nine inches, surveyed the house. He then nodded to the pianist. She began to play and soon the hall was filled with his voice, reaching into every recess and cranny. In his many rehearsals he had never sung it better. John, when eighty nine years old, said he could sing 'Holy City', 'as well as anybody.' The Bruces and the Ripkeys and his mates and their lady friends clapped until their hands were red and sore. No doubt his cheering section was the most enthusiastic in the hall. The judges announced that John Bruce was the winner. Mayor Gill of Melbourne presented John with a gold medal. One proud young lady sat in the audience and no doubt cheered the loudest. She was Miss Alice Ripkey.

Marsh and Company manufactured, under license, a machine to convert gasoline to burn in lights and lamps. The company installed and serviced the machine. In those days, when automobiles were few and rich men's toys, little was known about the danger of gasoline. Few knew of its flammability and how easily it could burst into a deathly ball of flame. It was more dangerous than people believed. One of Marsh and Company's customers asked that a repairman be sent to their residence, a house some fifty miles distant from Melbourne. John was assigned to travel there and make the repairs.

He rose early the morning he was to make the journey. He must get to the shop, assemble the tools and repair parts he may need, and get to the train station. The wagon driver helped John load and drove him to the railroad station. His gear safely on board the train, John sat back to relax and enjoy the trip. It was to be a pleasant break, a day in the country. It also was indicative of Mr. Marsh's confidence in John's rapid progress in his apprenticeship.

At his destination, a skinny, short gray-haired man met John with a one-horse small wagon. He introduced himself as 'Hap' and said, "I'm here to drive you to the ranch." Having made this announcement he helped John load his gear onto the cart. During the five-mile drive, 'Hap', having a new audience, regaled John with tales, legends and lies about the region and its people. He talked of squatters and their rights and others he called 'land grabbers', interlopers and less flattering names.

John was shown the gasoline conversion machine. He opened his toolbox and began examining the machine parts carefully. What John did he couldn't remember, or maybe never knew BUT suddenly he was ablaze, his body engulfed in fire. His screams brought back 'Hap' and three others. Being practical men they reacted promptly. One rushed to get a blanket. The others beat the flames down with their coats. When the man returned with a water soaked blanket they rolled John into it. The flames were out but not before John was badly burned.

He lay there suffering from third degree burns. One man carefully cut away all John's clothing. By this time, several others had gathered. An elderly, frumpy looking woman wearing a gray dress and white apron moved over. She shook her head before saying, "Load the wagon with straw and get this man into the doctor in town as fast as you can. Harry, you jump on a horse and make sure the doc is there when we get this man to his office." She went into the house and returned with some sheets. Then she poured some laudanum down John's throat. Four men gently lifted the fire victim on the straw, now covered with a mattress and a sheet. "Drive carefully. If he starts to scream too much, give him some more of this," she said as she handed the wagon driver the bottle of laudanum.

And so John began his long painful recovery from his third degree gasoline burns. After weeks in the hospital, he returned home to continue his convalescence. For a time, it was thought that he might lose an arm. Seven months later, the doctor told him he could return to work but could not go swimming for at least two years. John's career as a speed swimming competitor came to an abrupt end.

During his hospital stay his most frequent visitor was Alice Ripkey. The Bruces came, two or three each day. The Ripkey daughters often accompanied Alice. Amy Bruce never came. Each day she sent John some delicacy she had prepared for him, a piece of his favourite cake or pie, and the sugar cookies she thought he liked so much. He didn't. When he had recovered sufficiently John began to read. Alice borrowed books from everyone, every group she knew, for John to read. Mr. Marsh and 'Sandy' visited him at least once a week. Even the silent Mr. Plunket came several times. Each time he brought John butterscotch taffy. From somewhere, John never knew where, 'Sandy' found three books on sanitary engineering for the burn victim to read. John studied them, reread them, made copious notes, and was determined to learn more of this new and ever more important subject.

John returned to Marsh and Company to find the sheet metal shop staff had doubled in size. The work force now numbered eighteen. All were busy operating hand shears, rollers, edgers and seamers, Some were busy using soldering irons; all were engaged in manufacturing metal containers of various sizes. All this increased activity added to Mr. Plunket's workload.

Mr. Marsh blamed himself for John's accident. If asked why, it is doubtful he could tell anyone. Naturally, he was most happy to welcome Young John back. Noticing that John was not fully recovered he told him that "For the present, John you will 'help out' Mr. Plunket here in the shop." John nodded his head obligingly and Mr. Plunket smiled at the news he was to have an assistant. The fire burn victim was put to work helping to make up the work orders and parts lists. Within two weeks he was doing this entire task without help. He began assigning crews to different work sites. John had worked with all the journeymen at one time or other and knew who did what best. Using this knowledge he assigned the plumbers according to their special skills or lack of them.

When John saw a job that provided him with the training he needed he added himself to the crew as the apprentice but not before he had Mr. Plunket's permission. These jobs rarely lasted more that a day or two.

In this way John served out his apprenticeship, working in and out of the shop. If Mr. Marsh had believed in job titles, John may well have been classified as supervisor. His special relationship with Mr. Marsh in and out of the shop put John Bruce in a class by himself in Marsh and Company. He was also a close associate and the favourite of the clerk, Mr. Plunket.

When the South African War {Boer War} was declared in 1898, John offered to go. In the recruiting ceremonies, the junior and senior militia regiments were lined up on the Parade Square. All in their red jackets were marched around the parade grounds. John Bruce and his fellow engineers were taken aside where they were told, "we deeply appreciate your offering your services to go to South Africa but it has taken us years to train you men for the work that you're engaged in. For the Boer War, all we require is men who can be trained in five or six weeks and who can ride horses." They dismissed the regiment of engineers.

Pacifism was to John Bruce an abstract philosophy, at the same time an inherent belief. The War in South Africa was real. The British Government alleged the Boers {the Dutch settlers} were a grave danger to English interests. At first John was taken in by this clever, carefully-concocted propaganda. In dribs and drabs, he learned that Cape Colony and Chartered Company and the South Africa Company promoted the war. British business interests in South Africa saw the Boers as rivals, stealing land from the Blacks; natives who had occupied this region for generations. Cecil Rhodes and his fellow land pirates were out to protect their monopoly right to steal the land themselves. They had no intention of sharing it with the Boers. When John discovered the reasons for the war in South Africa he was extremely disillusioned.

"That really strengthened my feelings so that I started talking against the war." John said some seventy years later.

Foot Note One- *On April first, 1881, in the Colony of Victoria the eight-hour workday was proclaimed. The statute made the legal workday eight hours. This Australian colony was the first to set the regular workday at eight hours. William Bruce, John's grandfather, cut stone for the Burke Street cenotaph commemorating the passing of this legislation.*

Foot Note Two- On Labour Day, nineteen sixty-five, John Bruce was the first to receive the William Jenoves Award for his dedicated service to the labour movement. In accepting the award, he made one of his excellent speeches and tears came to his eyes and rolled down his cheeks. He was not acting. He was simply expressing his deepest being. As a duck takes to water, John Bruce took to elocution. To put it simply, 'he was a natural.' His simple beliefs, his strong convictions could be seen in his eyes, his face, his posture and heard in the words he spoke.

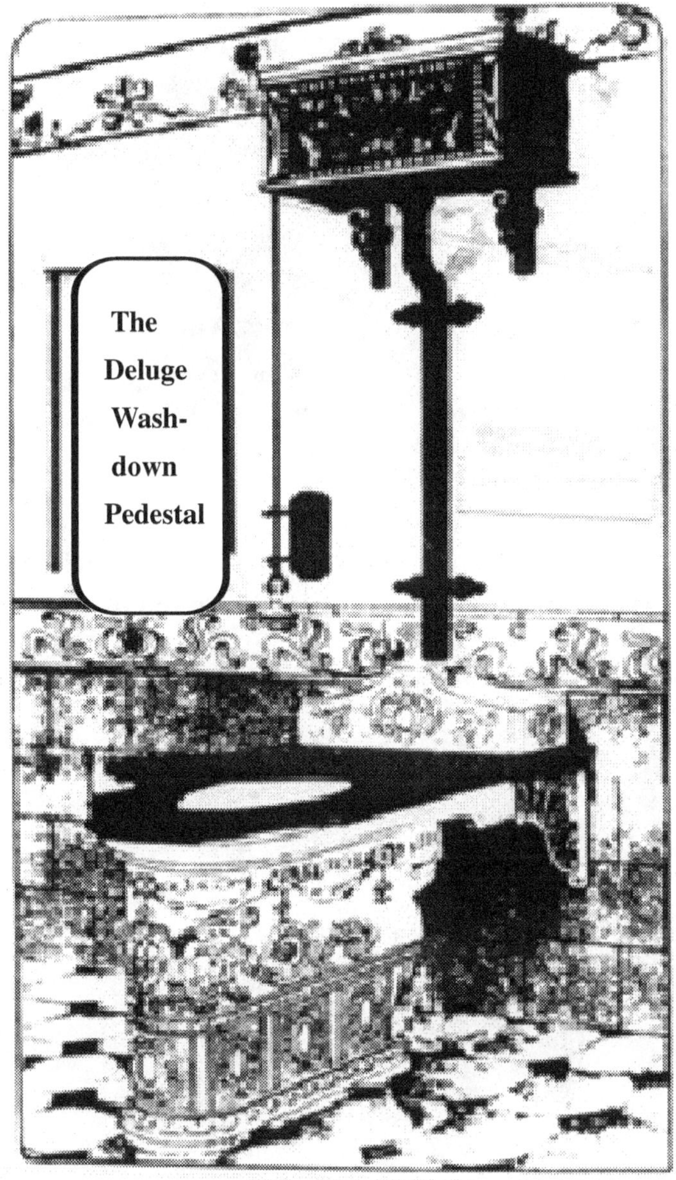

**An early Twyford W.C.
with siphonic cistern**

from the Thomas Crapper Catalog

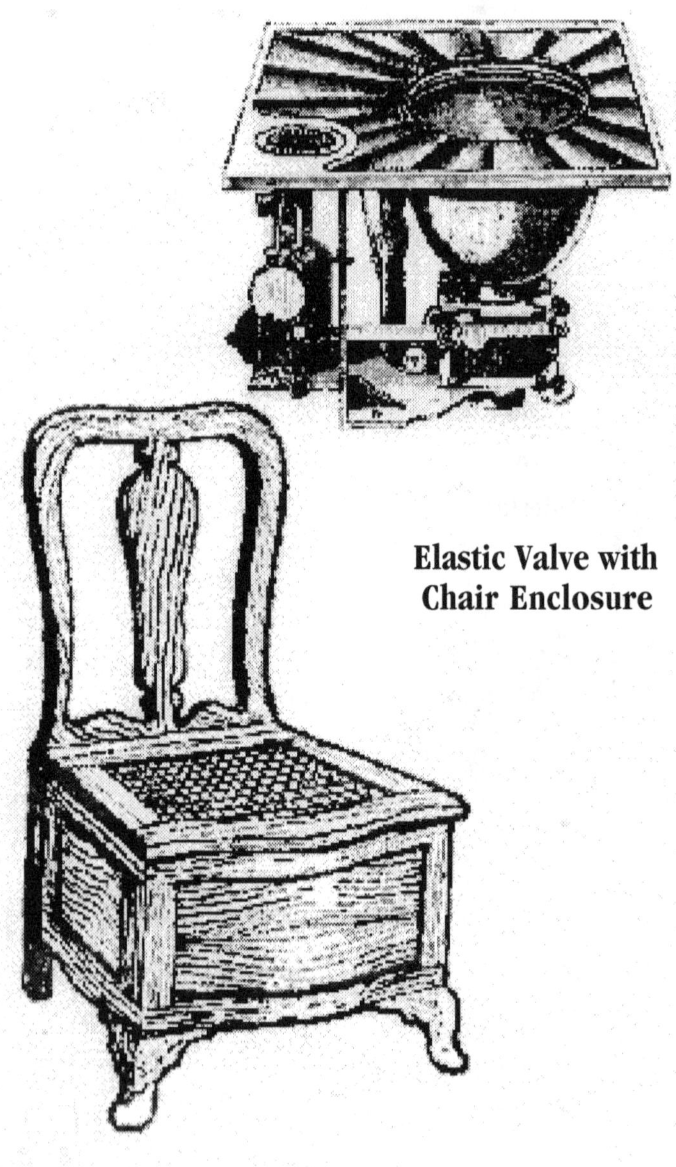

Elastic Valve with Chair Enclosure

from the Thomas Crapper Catalog

Pedestal Lion Closet 1896

The Sultan with decoration, 1896

Improved Aeneas with raised ornamentation

Dolphin 1880

from the Thomas Crapper Catalog

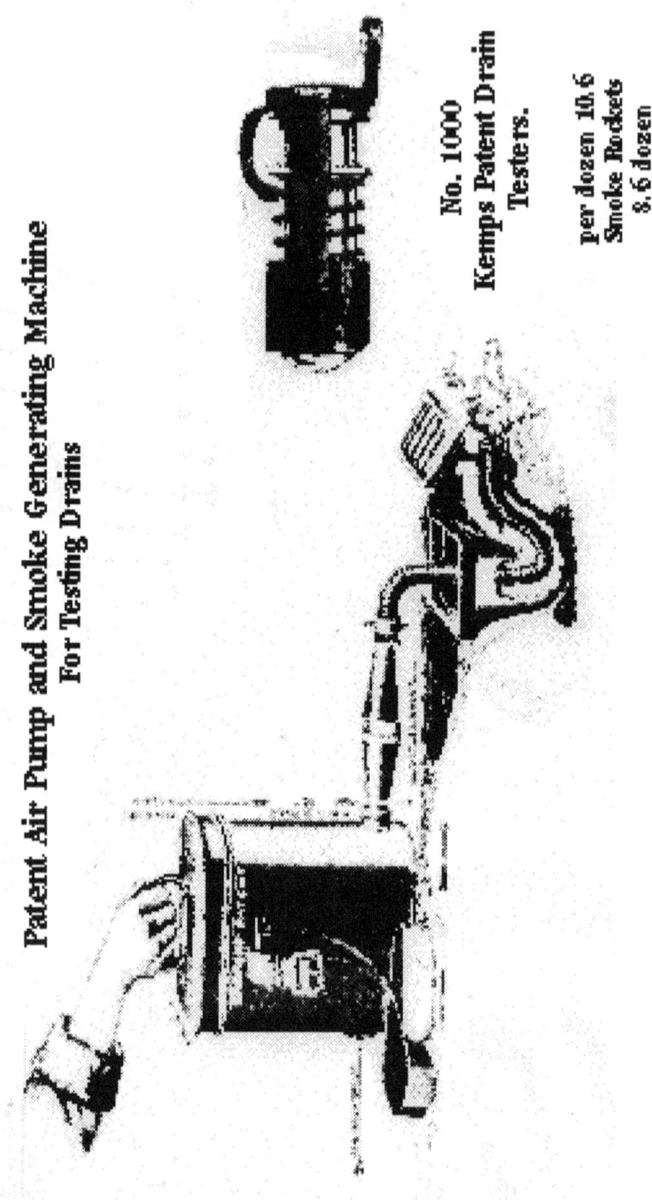

from the Thomas Crapper Catalog

Chapter Six - The Social Convener

By the age of eighteen, John Bruce had reached his full height of five foot ten inches. He was three to four inches taller than his fellows. Although not fully grown, he was broader in the shoulders and more solid of frame than the norm. He could out run, out swim his mates and was an expert rifle shot. He was earning a pittance as a second year apprentice, most of which he contributed to the family's meager reserves. His social activities, though limited filled many leisure hours. He sang in the choir; attended meetings of the Band of Hope, a youth temperance club. He belonged to the Presbyterian Church young people's group, and served in the militia as bugler and drummer before transferring to the army engineers. Still he had time, some evenings of the week, to join his mates in the village square. With little or no money to spend, they strolled about, stood on the corners of the village business centre. lounging about telling stories, ogling the young ladies walking by, greeting friends, neighbours and relatives. It was a small village, a little community where every one knew everyone and there were few secrets. Their activities, conversations and chatter were mostly repetitive. Each knew what the other was about to say. These idle evening hours John Bruce did not find very rewarding. He liked to keep his mind busy with new ideas, new thoughts, or be physically active. As other alternatives were few, he, like his fellows, walked, watched and exchanged comments, jokes and complaints.

By his nineteenth year, he began displaying another ability. He was able to see new possibilities and potential in events he was witnessing. He was able to look beyond the immediate; to visualize other concepts.

One day, in the second year of his apprenticeship, he was sent to do a plumbing repair job for Miss Skinner She owned and operated a dance studio that was located on a side street. It was a store located about two hundred yards away from the main thoroughfare on which Port Melbourne's stores and professional offices were situated. It had an unfortunate location and was an unwanted size. It was too big for a barber shop or tobacco shop, yet too small for a draper's or general merchant's purposes. Also it was located too far from the main business section to get any casual trade. Miss Skinner was the beneficiary of its original owner and builder's poor judgment. The ground floor was large enough for most of her dance classes. This floor was designed as a store. and as such, it had a high ceiling and heavy duty oak floor. She needed to spend little to convert it for her purposes. She had selected a pale blue paint, really a white with only a hint of blue in it, to decorate its walls. On the walls, she had pasted paper cutouts of the soles of men's and women's shoes. The right foot was a red and the left a green. These were arranged into the patterns of the various popular dance steps.

On the second floor of this two-story building, was a four-room apartment. Miss Skinner had attractively decorated and furnished this apartment. It was here that she entertained her friends and favourite students. She was not old enough to be an old maid but she would soon be classified as such.

John walked along Grant Street, the business section of Port Melbourne, turned down the side street and knocked on the dance studio door. Its owner welcomed him and told him what she wished repaired. As he busied himself with this simple repair job, while working he had an idea. Would it not be better to learn to dance properly than to stand about on the business sector on Grant Street at night, idling your time away? Not only would he and his mates learn a new social skill but meet some young ladies as well. Where else could young men in their late teens hold a young lady's hand, place one arm around her waist and have such behaviour accepted by polite society? Was it not a joy and a little exciting to hold an attractive lady dancing partner? John himself had an eye for the fairer sex. Such a possibility had pleasant potential.

John never hesitated to speak to his elders. He had no qualms in asking Miss Skinner if she would help to organize dancing classes for the young men of the village of Port Melbourne. Being interested in the young people in her small community, she agreed to hold classes. John had to recruit enough lads to be able to conduct proper classes. Miss Skinner wanted each male dance student to have a female partner to help him learn.

John had the bigger problem of persuading his male chauvinist mates that learning the fox trot, waltz and other dances popular in the 1890's would be fun. Some thought that if they took dancing lessons their mates would ridicule them. John had to use his considerable speaking ability, and quick mind to persuade his fellows to become members of Miss Skinner's dance class. No doubt John gave great emphasis to the presence of some of the village's finest young females as they would be serving as dancing instructors.

John Bruce and his friends showed up for Miss Skinner's first class. There, assembled in the dance studio. was the 'La Crème de la Crème" of Port Melbourne's embarrassed, unsure, curious working youth. Each of the heavy-footed, poorly coordinated recruits attempted to avoid notice. They all knew each other. There had been little or no personal contact between the sexes, save for a brother and sister who pretended, without success, not to know one another. Yet it was these two siblings who helped to dissipate the reserve and awkwardness that initially prevailed between the sexes. Each new male dance pupil was provided with a partner. Miss Skinner no doubt coached every young lady before the first class. She cautioned them not to laugh, or poke fun at their uncertain, untalented charges. Each young lady must with tender care coach, patiently encourage her pupil to try, and to persevere until he acquired sufficient skill to go on to learn another dance step. The wall displays were used to teach new students the rudiments of each dance. Each set of male and female foot prints on the wall illustrated the sequence of steps for a dance; be it a waltz, a polka, two step and other dances that were popular at that time. {The Sailors' Horn Pipe, etc., were dances taught her younger pupils.} Each young volunteer dance instructor used the wall display to introduce his or her partner to each new dance. John and his mates began their first lesson, each lady teacher stood at the side of her charge, facing in the same direction. Their instructor pointed to the wall at the foot pattern, then moved her feet to follow the pattern on the wall. As she did so she had her pupil follow along. First step, second step and so on. She led her partner through the dance pattern. She moved with grace and rhythm, hoping her awkward, uncertain partner would follow. On each appointed evening, the pianist banged away, all the young ladies and the chief instructor would patiently tolerate toes stepped on as their charges slowly acquired passable dancing skills. John Bruce was the first to actually dance with his partner in his arms, a goal to which all his mates now aspired to with renewed enthusiasm. They envied him this pleasure.

Miss Skinner ever watchful, ever helpful would assist a young volunteer teacher who was having difficulty with her partner who was not well coordinated. Certainly none of them were as well coordinated as John. With much effort and what seemed unlimited determination the participants laboured on. Thus they learned the waltz. The next dance, the quick step they learned with less difficulty. All the lads were anxious to acquire some skill and were tactfully encouraged to dance as well as they could.

John never left anything to chance. He kept constant watch on his lads. Any who needed encouragement, he helped. Those who did not fit in and complained, he tactfully persuaded to quit the class. He already had a potential recruit waiting.

He was always alert to Miss Skinner's beck and call. Alice Ripkey, John's most frequent dance partner, made mental note of John's careful monitoring activities. She began doing the same with the young ladies. She helped John play matchmaker.

Not all the young lads were Greek Adonises, nor were all the young ladies beauty queens. Not believing that opposites attract, John and Alice matched some males with new female partners, sometimes at their own request. John noticed that some of his mates, not excellent dancers, and somewhat less than good looking and not having all the social graces, became the regular partner of one of the more attractive, desirable ladies. In a few case the couples later married.

The task became less onerous, as personal relationships grew. In the beginning the young ladies had no special partners. As time passed, an instructor and a student sought each other out before the dancing began. Little by little, the couples paired off, as John Bruce did with Alice Ripkey. Whether it was her skill as an instructor or John's natural ability or both, he proved to be a better dancer than the others. Although John enjoyed dancing with other partners, Alice. with a watchful eye, limited his dancing partners to young ladies less attractive than she. Their contemporaries, who were not invited to join the dance class, made caustic and rude remarks at John and his mates as they entered Skinner's studio, wearing their Sunday best. The cat calls and unflattering comments ended when they saw dancing fellows dating the young ladies who had taught them.

It was a small village and soon everyone knew of the dancing class and each participant in it. Other young workers, learning of the classes, asked to join. John, with hesitation rejected 'would-be' womanizers and those who drank alcoholic beverages. Success was such that there was a waiting list as Miss Skinner's studio could only accommodate thirteen couples. Even that number posed a minor problem. Yet, no one complained about the lack of space. The couples were required to dance a little closer together. This did not seem to bother them. John, as was the case all his life, was in charge. He was the leader, organizer and administrator.

It was much more enjoyable and held more attractive possibilities than standing in the village square. Nights when there was no dancing class; two to three couples of dancers would take walks together or meet at one of the young ladies' homes.

By the time the students had completed their dance classes, John Bruce had seen a new possibility, a dance club. He discussed it with Alice. She talked with her lady friends, as John did with his mates. All agreed it was a great idea. However, a club must have a name. Here the persuasive talents of the ladies were apparent. It is extremely unlikely that a male put forward the name of Daffodil. If he had, his mates would have howled him down. Not so when one of the prettiest young women suggested the name. The other ladies immediately supported her. No male, anxious to please his special dancing partner dare object, least of all, John Bruce.

So began a weekly Saturday dance sponsored by the Daffodil Club. The couples danced to the tunes now long forgotten. Once the Club was formed, other couples who had never attended one of Miss Skinner's dance classes joined and became regular attendees. John, Alice and the other club officers rigidly maintained social decorum at all times. Of course, John insisted that no alcoholic beverage was ever served or consumed at the dances. At least none was with his knowledge. John did not trust God to monitor the dancers' behaviour. This task he took unto himself.

Many a relationship that began between instructor and pupil blossomed into romance and marriage. Not to escape such a happening was the man who conceived the idea of dance classes for his mates, John Bruce. Two years and a few months after the first dance class, on Christmas Eve, 1887, he and Alice Ripkey were married.

Still the Daffodil Club held its weekly dance. The newly wed Bruces continued to attend until he departed for South Africa in the spring of 1902. By this time, Alice had given birth to two children and her love of dancing continued. John Bruce himself enjoyed dancing all his life and rarely missed the opportunity to trip the light fantastic. How long the Daffodil Club continued to operate after the Bruces' departure to Durban is not known.

Chapter Seven
The Journeyman Plumber

In 1895, John Bruce became a journeyman plumber and immediately joined the Plumbers' Union. From the beginning, he was an active member. The union membership meetings were held each week. 'This was one activity, which filled our hours." he explained. As if John with his many interests, need worry about filling his time.

His grandfather Taylor who had lived in trying circumstances in England had come to Australia in 1855. He had been a farm labourer and teamster before immigrating to Port Melbourne. There he secured employment as a wagon driver. Later he became a stevedore. That same year the dockers formed their union. John Taylor was one of the charter members. John remembered his grandfather Taylor being on strike in 1891. Later that year, grandmother Taylor died and the widower moved in with his daughter and her family. Thirteen Bruces already occupied the house.

John, now an accredited journeyman plumber, was not satisfied. When he began his apprenticeship outdoor plumbing was by degrees being replaced by indoor plumbing, an infant industry. His many days with Sandy repairing foul smelling polluted drains lived on to haunt him. It was still for the most part an unsolved problem. He believed the solution lay in sanitary engineering. John was determined to learn more about this subject.

He learned that Thurmond Vocational College, a technical school that had night classes that young workers could attend to learn new skills, was offering a course on sanitary engineering. The classes were to be held in the evenings. John enrolled in the course.

He found the classes were to up-grade a plumber's knowledge of the trade. The practical side of sanitary engineering was emphasized. The industry had the task of ensuring that every member of the community had his full supply of uncontaminated air, pure water and freedom from polluted drainage systems. The teacher said there were still far too many contaminated cesspools and poorly maintained private drainage systems. They must be eliminated and replaced with a modern sewerage disposal system.

The technology was available. The products were available. His purpose was to show the students how to use them; how to install the plumbing fixtures, pipes and drains as specified by sanitary engineers.

The instructor distributed the most recent plumbing catalogues. Among them was Thos. Crapper's. His patented products included such items as 'Seat Action Automatic Flush'. the W.C. cistern, the 'Valveless Water Preventer', the 'Trough Closet', a long bench like device with seats for eight or more to use at the same time. The Trough Closet was just that, a trough, over which was a bench made of a row of eight, ten or twelve toilet seats. The water entered the upper end ran down the trough. It was a continuous run of water. The stream of water carried the droppings away. Practical jokers would crumble up a wad of paper, light it, drop in the upper end and the flow of water would do the rest. Some thought it amusing to watch the victim jump as the burning waste drifted down with water flow to come in contact with the employee's exposed ass. Thomas Crapper redesigned the Trough and that ended the fun of the practical jokers.

The teacher would say, "On page ten, you will see a drawing of a 'Climax Release Valve'. This is it." He would hold it up for all to see. Then the class would gather around him, and he would show them how it was to be installed. Its threads were worn by the many times it was attached and detached. It served as a learning tool. The instructor knew what was up-to date and available, more so than most plumbing contractors, including Marsh and Company. He would use a glass beaker, funnel with a shut of cock and 'S' shaped glass rod to demonstrate how a common sink or drain trap worked under pressure.

During the course, the instructor went page by page through the Thos. Crapper catalogue and a few others. He said he used the Crapper catalogue because he had invented more plumbing fixtures, attachments, valves. cocks and drain accessories than anyone else.

He warned his students of the tendency to use larger drain-pipes than necessary. This belief that bigger is better is not usually true in building drains, he said. Larger drains cost more and required more cleaning. This added to the maintenance cost. He talked about the rate of fall for drainage pipes of different sizes. He distributed a chart used by the city of Brooklyn. He said he, himself, thought the rate of fall should be ten to fifteen percent greater than those listed. The length of the drain and the hardness of the soil often were reflected in the fall levels selected. He stressed the good ventilation of drains and showed them examples of ventilating pipe, elbows, and tees available.

A technical school instructor recommended that John enroll in Dr. Anderson's night school class on Sanitary Engineering at the University of Melbourne. John did. Here he hoped to learn the theory and technology he felt he sorely lacked. While convalescing, he had read and reread the somewhat dated books on sanitary engineering; the three books his co-worker, Sandy, had borrowed for him to read.

John was a problem student. He knew more about the subject than the other students did. He frequently asked questions that would take the professor off on a tangent. Complexities, although small, were ones that most, if not all his fellow students could not comprehend. John would nod to acknowledge that he understood an answer. Dr. Anderson realized the other students had not understood. Most of John's questions were to update and add to his knowledge of a subject.

Dr. Anderson emphasized that sanitary engineering dealt with the promotion of public health, comfort and control of the environment. His course dealt specifically with water supply, sewage, stream and water pollution, liquid waste and industrial waste. His lectures covered such subjects as engineering, a smattering of biology, bacteriology, chemistry, hydraulics and disease control methods. He assured his class that his lectures would not make them sanitary engineers but would make them aware of what they must learn to become one.

His lectures fell into these categories: public water supply, sewage, streams and water abatement. sewerage treatment, industrial waste treatment, refuse collection and disposal. During his talks he kept reminding his listeners that politics and public pressure could be a friend or foe of a municipal sanitary engineer.

He gave several frightening examples of stream pollution. Two that John remembered years later were: "In 1858, when Queen Victoria and her husband took a pleasure trip in a launch on the Thames River, the foul, rancid and putrid smell made them leave the vessel in great haste and hide indoors behind closed windows. [Little wonder that Queen Victoria's husband, the Prince Consort, died of typhoid in 1861]

That summer the heat wave exposed rotten garbage and human excrement dumped into the river. It lay along both banks. A river should never be used for such purpose. Members of Parliament adjourned early in 1858 to avoid the horrid, intolerable smell that seeped in to fill all the rooms of the House of Parliament at Westminster with terrible, nauseating odours.

The one tale that John enjoyed best was another story involving Queen Victoria. On her visit to Trinity, one of the colleges in the City of Cambridge, she and Dr. Whewell stood on a bridge overlooking the Cam River. Queen Victoria asked, "What are all those pieces of white paper [toilet tissue] floating down the river?" The Master of Trinity College said, "Madam, those are notices that swimming in the river is forbidden."

The mistake of using streams and rivers as disposal areas for waste was a common one. Dr. Anderson said that even worse results would occur should the winding, slow moving Yarra River be used for such a purpose.

In one of the books John read while recovering from his third degree burns it was reported that almost a quarter of the homes in London emptied their liquid waste into the street. Others used any ditch, stream, or river nearby as a disposal area for refuse from sinks, tubs and toilets.. In 1844 at Windsor castle, fifty or more cesspools constantly seeped out covering lawns and driveways.

Dr. Anderson regaled his class with horror stories of pollution causing serious health problems. For these problems, only good, advanced sanitary engineering could remedy.

Much of John's knowledge of sanitary engineering was dated; still it was more than his fellow students had. The books John read advocated the use of rivers and streams as disposal areas for liquid waste of all kinds. The new technology advocated by the professor called for holding basins. The solids in the liquid waste could settle and be dealt with separately in these areas.

During the first few lectures, Dr. Anderson had the impression that he was teaching only one student, John Bruce. The others behaved like mere spectators. His practice was to allot specific assignments to each student. John's would come back promptly, in more detail and more lengthy than the others. Dr. Anderson asked John to serve 'as kind of an assistant.' The doctor had two objectives in mind. One was to keep John from monopolizing the class time, and the other was to help the slower learning students. That year there seemed to be more of the slow learners than the norm. Maybe that was because Dr. Anderson was comparing that year's class with better informed, quick-witted, quick-tongued John Bruce. John was pleased to accept his new role. The two objectives were reached.

John was the only journeyman plumber in the class. He understood the practical application of sanitary engineering. He could picture the drains, valves, and holding basins the design called for. This impressed Dr. Anderson. He stored this knowledge away for his future use.

The new Utopia where all who wished to work had jobs and never went hungry was to have its dark side.

In 1897 and 1898, Australia suffered its second depression in a decade. The havoc of 1882 saw seven banks collapse and leave the country's economy in a precarious state. Their recovery was slow and uneven. Although the second recession was not as severe as the first -- still it brought the economy almost to a halt.

People had less to spend; retail sales fell off drastically. Factories laid off workers. Some closed. The boot factory sent no precut leather to David Bruce to sew and glue together. He had a wife and seven children to feed, so the Bruce's meager 'nest-egg' was soon gone. There was no money. Grandfather Bruce was also a victim of the depression. He worked briefly from time to time, so could do little to help his son, David, and his family. As fewer ships came and left the harbour, Grandfather Taylor, a stevedore, was only working a day or two a week. John could contribute little as he had himself and a pregnant wife to support.

There were no government welfare programs nor unemployment insurance for the unemployed to fall back on in the 1890's. The David Bruce family was without food. Each day, more and more people were starving, so the town council of Port Melbourne decided to use some of its limited budget to provide soup and bread to those who were without food. Funds were diverted, contrary to the municipal council's mandate. Funds originally appropriated for other purposes were used to set up a soup kitchen. To David Bruce it was charity, nothing but a hand-out to the destitute. Those in dire need could get a ration of soup and bread. David Bruce, with his children suffering from hunger had no choice; he must go to the town hall to get food.

John's younger brother, William, who stood in a corner of the kitchen to witness the events of this sad day reported it to his older sibling. This is how John described what he had been told more than a half century later. In the centre of the kitchen, his kind, gentle father stood, the look of hopelessness and defeat written in large letters on his face. The man was still trying to cope with an unacceptable fact. His wife Amy, ever supportive of him, was doing her best to subtly comfort him She put the handle of a large pot in his hand, a utensil in which he was to carry home the soup. Then she went to a shelf in the pantry and returned with an empty flour sack. This was to carry the bread in. All this the young Bruces watched, at first not believing what was happening. The reality came as their father became more depressed moment by moment. They had never seen their father with such a hopeless, forlorn look on his face, in his eyes. The children knew their father was devastated.

David turned and walked quickly to the outside kitchen door, the empty flour sack tucked into the large pot he carried in his left hand. As he passed Amy she touched his shoulder, tried to comfort him with her eyes, and brushed his cheek with a loving kiss. The door closed behind him. Amy lifted the corner of her worn apron to wipe tears from her eyes.

David's quick steps moved him towards the town centre. As he came in sight of the town hall, his pace slowed almost to a halt. Then with renewed courage he moved into the long line of men who were already assembled at the rear of the municipal building, each with a pot or a pail held in one hand. Their heads were down; their eyes riveted on the back of the man ahead. Many of them were on a first-name basis and knew the others by sight. Yet each remained silent, recognizing no one around them. Each hoping he, himself, would not be noticed. They would shuffle ahead as one of their number left, his pot or pail full of soup and with a few loaves of bread in his bag or sack.

The two town employees doling out the soup and bread spoke to no one; recognized no one; their eyes looking down at the task they were performing. They, too, shared their fellow workers' shame and humiliation. David, in what seemed to him to be a decade, finally reached the table that blocked the doorway. He tentatively held out his pot. It was filled with soup, his empty flour sack stuffed with four loaves of bread. David did not look at the soup. He knew it would be mutton soup made from old sheep. Such ancient animals were cheap; no doubt bought from some rancher for a few shillings.

Amy and the children waited forlornly for David to return. Almost an hour passed, when their father opened the kitchen door. To the family their father looked a man who had fought the demon bravely and had barely survived. David crossed the room, placed the sack with its bread and pot with its soup on the wood table. He pulled back a chair, sat, put his arms on the table top, his head on his arms and cried. John's younger brother strained his ears and he heard his father's almost noiseless sobs. Tears came to all the children's sad eyes and their cries were as quiet as their father's. David was a man terribly humiliated, almost totally defeated. All this, John's younger brother had seen in the few brief seconds it took his father to sit down at the table. In his heart, young William shared his father's mortification. After hearing of this sad happening, John knew this devastating experience would remain with him all his life.

The town council had no mandate to use town taxes to provide even this very limited relief to the hungry. The town fathers were aware of the power of more wealthy citizens. These pillars of the community might accuse them of misuse of the very limited town tax revenue. To ward off such criticism that no one should get something for nothing, these elected officials decided that each male recipient of the mutton soup and bread must do some work for the town.

A century later, politicians would call it 'Workfare'. Widows were not asked to meet this work requirement. Each man who had accepted the soup and bread handout was required to smash a half-ton of rocks into gravel. Half-ton piles of stone were dumped at the roadside. Each man had his own pile of rocks. He was given a 'knapping' hammer. There they would sit, smashing away, making gravel for the roadbed beside which they sat. This slow, laborious work was performed in the blazing sun. Sitting, they sweated at the roadside for all to see. The work could have been done in a stone quarry, with tools more appropriate to the task. Work done in a stone yard would not be seen by the people of Port Melbourne. The town councilors were above all else politicians. It is possible they may have regretted imposing such an unreasonable requirement upon those accepting relief, but not enough to save the rock smashers from humiliation.

It was slow work; done by the men who were not given the proper tools for such a task. It was demeaning for men to be required to break stone in this unproductive way. Each man began to smash rock, each doing so to his own time. As if prearranged, the hammers began to beat in unison. One man began to hum a monotonous dirge, a wordless moan without a beat. Others joined in a chorus. No one looked up, or indicated they heard this song of mourning.

John was one of the three James Marsh's employees traveling in a horse and wagon to a new work site. As they moved down the road, John saw dozens of men seated at the roadside. Each had a pile of rocks before him. They were using a knapping hammer to make little rocks out of big ones. John saw the father whom he loved and admired sitting among this work for welfare crew. He saw him sitting at the road side suffering the summer heat and humiliation. It broke young John's heart. How, he asked himself, could the town council be so heartless, so cruel? Why should these unemployed be subjected to such public humiliation, their poverty exposed for all to see? Why must a man's self-esteem be shattered because he could find no work? This event strengthened John's beliefs in the merits of socialism as expounded by Karl Marx. Such thoughts as these passed through John's mind as he stared at the floor of the wagon directly at his feet. He did not wish to see his father or have his father see him witness his humiliation.

Seeing his kind-hearted father, so inhumanely treated, made a deep impression on young John. It did much to strengthen his belief in the right of all men to live in dignity and freedom; those freedoms so graphically depicted in the paintings of Norman Rockwell's 'Four Freedoms'.

The vision of his father's humiliation lived with John Bruce all his life. Three quarters of a century later, when he told of his father smashing rocks at the road side, the bitterness, the shame of it all, was as vivid to him as if it had happened only yesterday.

Ever after that, John would act with haste to prevent anyone from being humiliated nor would he permit himself to be humiliated. His quick, aggressive response to a person, who talked down to him, earned him a reputation as being a man with a short fuse. He could also be quick to anger for other reasons as well, but never as quick as he was to prevent someone, himself included, from being humiliated. A few such episodes are recorded in this book as proof of the above statement.

Another Marsh employee, Bill Taylor, who worked under John's supervision, laid charges against him with the plumbers' union. He claimed John, who was a foreman, should not be allowed to keep his union membership. Bruce's attitude over the years shows that despite what he said, he never thought very highly of Englishmen. Taylor was a "Limey". He had taken on trouble when he laid charges against a Bruce. John, ever a militant, was highly regarded by his fellow unionists. Taylor's charges never had a hope. They were dismissed almost immediately.

In 1897, Melbourne decided to install sewers to link up to indoor plumbing. This meant building a massive sewage system throughout the city. Plumbers were in great demand. Melbourne posted a notice of an opening for the position of Chief Plumbing Inspector. John Bruce took a trade test to qualify for this new and special work. His night school classes better fitted him for the job as advertised than most of the applicants. He was now 21 years old. Each applicant had to write an exam. John Bruce took the examination. His marks were the highest of all that had taken the test.

Despite this he did not get the job. John, always quick to take offense, reacted as one would expect. He rushed into the office of the Chief of the Metropolitan Board, Mr. Copp, demanding an explanation. The Chief told John, "You are too young for the job. You would be in charge of men at least twice your age, who are plumbers with more years of experience than you". This explanation only temporarily satisfied Bruce. Then he learned of the fate of Jimmy Duncan. This Scot in his mid-forties had received the second highest marks in the examination. He was near twice John's age. The same argument could not be used in his case. But Jimmy Duncan did not get the appointment.

The job went to a "cheap politician and a non-union man". The plumbers of Melbourne knew him as a Conservative hack that didn't have the qualifications or practical experience to do the job. (John's words) John thought his employer, Mr. James Marsh, may have had something to do with him not getting the post of Chief Plumbing Inspector. His employer thought very highly of him and didn't want to lose him as an employee. He treated John very fairly. John said that Mr.Marsh treated him as if he were one of his sons. The proprietor of Marsh and Sons selected John to supervise the shop. Despite John's youth, he was given the responsibility of running it.

John was bitter about his failure to get the position of Chief Plumbing Inspector. The depressions of 1897 and 1898 only added to his loss of faith in the future. He still had 'itchy' feet.

So, in the winter of 1902, he welcomed the job opportunity of becoming Chief Plumbing Inspector for the City of Johannesburg. S. A. that was offered him by Dr. Anderson. He discussed moving to Johannesburg, South Africa with his wife. Alice. "She showed no opposition to me going," he reported. It is likely her willingness to move was that she had seen how disappointed her husband had been. She was witness to his obvious remorse when he failed to get the position of plumbing inspector with the City of Melbourne.

Foot Note

From The Review of Reviews
Vol.VII No.41
New York, June, 1893

The Australian Collapse Was Unavoidable
The bursting of that bubble sooner or later was simply inevitable; and to have comprehended the story of Argentina. Inflation and collapse would make it comparatively easy to understand the terribly drastic process of liquidation that is now going on in Australia.
Nearly every bank of any consequence in that New World has been obliged to shut its doors within the past two months, every commercial interest has been involved. With the bank in the deepest distress. Australia has as yet a small population, and its legitimate annual wealth production amounts to a very modest total. The Colonial Governments have far overstrained their credit by incurring heavy indebtedness for all kinds of public enterprises many of which could not become productive for a long time. Australia has been living at too high a rate and spending beyond its means. Its banks were

largely the ministers to this extravagance, for they succeeded by means of numerous branches in England and particularly in Scotland, in securing enormous deposits through the high rates they paid; and this money was far too freely and carelessly loaned in Australia for speculative and unsubstantial enterprises. Finally the colonies had reached the extreme limit of their ability to borrow. Whether for public or private uses, and the English and Scotch depositors had taken alarm and begun to call in their money. This was the beginning of the end? The money had been loaned to people who could not pay it back except by borrowing it elsewhere, and the sharp annihilation of credit caused a complete collapse.

Chapter Eight - South Africa

In the last decade of the 19th century, several hundred Australian building tradesmen migrated to South Africa. They and their wives sent back glowing reports about their life there. Although they didn't say it in so many words, they hinted that the 'streets were paved with gold.' The recipients of the letters were impressed. Moving to South Africa was often the talk of John Bruce and his 'mates' in Joe Eaton's barber shop.

Dr. Anderson, a professor at the University of Melbourne, taught sanitary engineering. John Bruce had been one of his more talented students. When the City of Johannesburg offered the professor the position of Chief Medical Officer of health, he accepted. He asked John Bruce to join him and be Chief Plumbing Inspector. John had failed in his attempt to get a similar position in the City of Melbourne. He saw this as a chance to set the record straight.

Some of his fellow plumbers and other building tradesmen he knew were going to South Africa. He had always had the wish "to see the world". This would be the beginning of his journey, In May, 1902, John Bruce and some of his colleagues sailed for South Africa. Alice and his two children were to join him later.

John Bruce looked at the overcast sky as he walked up the gangplank. Storm clouds hovered on the distant horizon. This made him a little apprehensive for he was not a good sailor. All his life he had a cast iron stomach but only when he remained on dry land. The safe harbour of Port Melbourne was no problem for him. It was the turbulent deep water that lay out beyond Port Phillip that worried him. Here the ocean could be rough, with great rolling waves that could turn John's stomach upside down.

He looked forward with great expectation to what he thought awaited him in South Africa. He was uneasy about the South Pacific waters he must travel to reach there. It was a subdued John who climbed into his bunk that night. His evening meal, although it looked appetizing, he hardly touched his food. The less he ate, he thought, the less the risk. The gentle rolling of the ship lulled him into a dreamless sleep. When he awoke next morning he looked out the porthole to see a bright sunlit sky and a postcard sea.

En route John Bruce and his five "mates' explored the ship. They walked about the deck and had polite conversations with their fellow passengers. By mid-day they were full of the holiday spirit. John, always a good mixer and easy to meet, was soon leading the singing and having a great time. The six Aussies gave up trying to button-hole those who could tell them about South Africa. It was a holiday and one to enjoy .

The radio operator received word that a treaty had been signed with the Boers at Vereeniging. The South African war was over. The struggle that had brought great prosperity to the whites in the colony had ended. It was good news. The passengers and crew cheered when they heard it. The war's end would change much in South Africa and John Bruce's plans. An estimated 60,000 Boers had embarrassed the British with their tenacity. It had taken 448,435 English and Commonwealth troops and two-and-a-half years to win the struggle. The British were out-generaled by the Afrikaans. Lord Roberts, a career soldier of the old school and Lord Kitchener, an ambitious, posturing officer were heralded as great military heroes. The English enjoyed creating heroes out of men wearing gold braid.

The Boers had taught the British, Roberts and Kitchener included, the art of guerrilla warfare. The Brits learned the hard way that it was easier to see and shoot someone wearing a red jacket than one in a drab khaki coat. In keeping with tradition the British generals had blundered through to win the final battle.

When the ship docked in Durban John said good-bye to his fellow passengers and his five "mates". They remained in the port city. They had no problem finding employment in South Africa's second largest port city. He was sad when he boarded the train for Johannesburg alone.

Going to South Africa was not one of John's smart moves. It was a mistake he no doubt much regretted, although he never admitted it.

South Africa was a land stolen by the English and the Dutch from its black inhabitants. No need for bravery, when you had guns, cannons and all the ammunition you required and your enemies had only spears. In 1893, Frank Edwards, a journalist, wrote, "Speaking of Mashonland {then occupied for decades by the Matabeles} on the whole, I should say its one of the richest countries added to the British colony, built on extravagant expectations of picking gold out of the ground by the spadeful."

From a safe distance, the English and Dutch could kill the militant Matabeles with ease. No fool, their chief Lobengula evacuated much of his land and the English and Dutch moved in. The exploitation and murder of natives was condoned by many including the Rev. C. Carnegie, a man drunk on the gospel of greed and suffocated by the fuzz of faith. He may not have spoken for all the churches when he wrote a piece entitled, 'Among the Matebeles' which was published in the magazine, 'Sunday at Home'. In this article, he condemned Chief Lobengula for "cruelty, butchering and misery"; the same crimes the Dutch and English had and were committing.

John Bruce saw Cecil Rhodes and Leander Starr Jameson, as the instigators of the invasion and slaughter. They were praised, not condemned, although the land pirates and killers of blacks, that they were. This John could not understand. John saw it for what it was; genocide. The number of natives killed or driven off the land, upon which they had lived, appeared not to be worthy of any newspaper reporter's attention. Better the public not know, so it was not reported.

John knew that Rhodes, Jameson, the Cape Colony and Chartered Company and the South Africa Company committed murder. Nefarious acts were ignored or condoned by the Gladstone government in England. John was appalled at the silence of the Christian missionaries. They acted as allies of these land thieves.

Among the church converts was a Khama, Chief of the docile, subservient Mashonas. This Christian "brain washed" black, wearing his three-piece English-tailored suit, became the vassal of the English. His unwarlike tribesman marched with the English mercenaries to drive the Matabeles off the land. In 1893, Mr. J. Theodore Bent said in *the Contemporary Review*, that he held "no prospect of the Mashonas or even Khama's men proving much use as fighting allies."

The Matabeles were a breakaway group of Zulus. They like the Zulus, were the Prussians of Africa. In the above mentioned article, J. Theodore Bent said, in South Africa, "there is no tribe which can stand up to the Zulus." The Matabeles fighting for their land were braver than the English greedy hired guns. If Chief Lobengula and his 20,000-man army had been as well armed as the hired soldiers of the Cape Colony and Chartered Company and the South Africa Company, the history of South Africa would have been much different. Spears and leather shields are no matches for pistols, rifles and Gatling guns.

When Rhodes, Jameson, and their fellow land pirates required help to steal the Matabeles' land, it was there. The British government in England sent in troops and police. Their ultimate crime was to appoint Rhodes as the Governor of the Cape Colony and to knight Jameson. The English won easily their war with the Matabeles. Each self-serving mercenary was given 3,000 morgens {6,000 acres} of stolen land.

All during this horrible murderous theft, missionaries throughout Africa kept busy converting the black Africans to Christianity. Yet the white God-intoxicated bible thumpers with very few exceptions did nothing to curb or speak out against the murdering, land stealing of the merciless brigands. Black converts to Christianity were not allowed to sit in the same church as the whites. Blacks were not allowed to sleep over night in the same dwelling as a white. The British version of apartheid was as brutal, as rigidly enforced, as that of the Afrikaans several decades later. The British churches condoned apartheid. After South Africa became independent they condemned it.

The British land pirates were following the example set by the Dutch a few decades earlier. Led by Andries Henrick Potgieter, Andries Wilhelmus Pretorius and others the Dutch inflicted heavy casualties on the Matabeles and stole their land. Guns and gunpowder determined the winner of this war. The Boers had what the English wanted. The British did as they had always done. Send in the armed forces to take possession of the land that belonged to others.

The self-serving English, the Colonial Office and the Gladstone Government grossly violated all the ethics of John Bruce, young recent arrival in the Cape Colony, and he could not sit silently by. Typical of John, he began speaking out against apartheid and the cruel abuse of the Blacks.

In 1980, Margaret Thatcher, the British Prime Minister, following the example of her predecessors supported the white supremacists in South Africa. In July 1995, she ate humble pie as she sat as a visitor in Westminister Hall and listened to Nelson Mandela. A man she had branded as a leader of a terrorist organization. He reminded the British that they were the fathers of apartheid. The Afrikaans who came after them were just carrying on a British Tradition.

A clipping from the Toronto Globe and Mail

Mandela Chides Host for Colonial Past

London - Britain heard a different Nelson Mandela yesterday as the South African President in a rare address to Parliament, chided his hosts for their colonial past and made a somber appeal to the world not to neglect Africa.

Mr. Mandela said his presence in London as a head of a multiracial democracy marked the 'closing of a circle' that was two centuries in the drawing.

But for the first time in his state visit, Mr. Mandela gently but firmly reminded Britons that it was their colonization in the 18th century that sowed the seeds of white supremacy in South Africa.

Mr. Mandela was the first foreign statesman to be accorded the honour of giving an address in Westminister Hall since France's Charles de Gaulle in April of 1960.

Former Conservative Prime Minister Margaret Thatcher, who once described Mr. Mandela as a leader of a terrorist organization and stubbornly refused to join in international sanctions against South Africa's white regime in this 1980's, attended the speech. —Reuters

The original inhabitants were treated no better than farm animals and at times not as well. Unfortunately, few voices joined John in protesting these gross injustices. He considered the British behaviour to be totally unjustifiable and unacceptable. It was contrary to his entire religious upbringing.

This was the land in which John Bruce arrived in early June , 1902.

He had been taught that all men were equal. With his ready tongue, he was sooner or later to anger the powerful perpetrators of this villainy.

Shortly after the two-and-a-half year South African War between Great Britain and the African government of Transaal and the Orange Free State ended, the British government demanded Johannesburg pay ten million pounds as its share of the cost of the war with the Boers. 'That was a great deal of money in those days, a pound being worth five dollars Canadian." John said: "Typical of the British, the Boers were to be bled." The French complained that, "The British will fight to the last Frenchmen." This could be paraphrased to 'Let the Boers pay". The victorious British demand for their pound of flesh bankrupted the economy of Transvaal and the Orange Free State. The English were to do the same to Germany in 1919, after World War One. This vengeful act destroyed the economy in South Africa. After World War One, the Brits did it again; created the economic chaos in Germany that set the stage for Adolph Hitler's rise to power.

By this time, Cecil Rhodes, the king of the land pirates, was dead. (March 26th, 1902) His long-time fellow land pirate and murderer Sir Leader Starr Jameson and his ilk, escaped paying for the war they had provoked. This appropriation spread to bankrupt South Africa. "Everything went dead" John Bruce remembered.

The city had no funds to pay Dr. Anderson who had come to serve as its Chief Medical Officer of Health, or John, who was to be the Chief Plumbing Inspector. Johannesburg gave each of them "three months employment to carry them over until they got on their feet". Gladstone was insistent that others pay for his war.

Mrs. Bruce thought her husband was an attractive man. She knew other women thought so too. John enjoyed the presence of pretty women. His good singing voice and his ability to dance well made him quite popular at social functions. She thought that women he met at church were less likely to cause problems for her. It was better that his social life in South Africa be within the church environment.

Taking long strides, John took off for the Presbyterian church to comply with Alice's wishes. As he walked he thought of all the black African converts his donations to church mission funds had made possible. He wondered what his grandfather, William, would think of the kirk of the Church of Scotland in Ardie if the pews were half filled with blacks. He would write home that night and tell about his first visit to a church in South Africa. Up the steps he went. Upon entering the church he looked around and saw no blacks, not one. As John looked about he saw the private pews reserved for those who could afford to pay for them. He found his way to the public benches and took a seat. He barely had time to study the structure before a haughty middle-aged man came up to him and said, " You can't sit there."

"Why" asked John, "these are pews for the congregation?"

"Yes" said the sexton, "but you will have to wait until all the regular parishioners are seated. If there is one vacant you can sit down". No welcome for John here, only a reprimand.

John was so angry, yet he paused a moment to curb his quick tongue. Bruce took an envelope from his pocket and shoved it into the sexton's hands. With great effort controlling his anger, he said, "This is a letter from my minister in Australia. I came here hoping to worship in this church and find some opportunity to be of service but you can tell your minister for me this is the last time I'll ever darken the doors of this church or any other church again."

In reporting this incident years later John said, "At that moment all my Christian values went from me. My church was just another creature of the rich and privileged". The doors of this church [the building] were not open to the blacks." His recently-abandoned faith did not believe in the ' Brotherhood of Man". John did not believe in the "Brotherhood of White Men ONLY". When Bruce was eighty-seven years old he said, " I am as big an agnostic as anyone one can be." He never again "darkened the doors of any church" except to attend a wedding or a funeral. John wrote his wife that night but did not mention his visit to the church.

John Bruce was a disturbed man as he set out for his temporary job Monday morning. All night he had fumed about the treatment afforded him by the church.

The mountain air did little to cool Bruce down; his anger remained with him. The terrible conditions of the blacks, the native converts, his former church accepted without protest. The more John saw, the more firmly he was convinced that his quitting the church had been a wise decision. It was the only one he could have made. Two thirds of the whites were British nationalists. This was an English form of apartheid and Bruce didn't like it. His job with the City of Johannesburg finished, he found other employment. He and his mate and brother-in-law, Bill Hatfield went to work building the Cape Town to Cairo railway. It was to be cut through the jungle, joining the two most distant cities in Africa. The real purpose of the railway was to link up the lush farm lands and gold and diamond mines of Transvaal and Natal with a sea port. This area later become Southern Rhodesia, a separate colony named after the English brigand. The idea of linking Cairo with Cape Town was to cover over the duplicity of the land pirates. This railway never linked up the two cities nor was it ever intended to do so. Shipping by water between Cape Town and Cairo was cheaper and more practical.

John, with his colleague, Bill Hatfield, set out for the job in Bulawayo. A man might be proud to have played a part in building such a significant venture. The tracks were run through the jungle which had its attractions and its dangers. Bruce and Hatfield had worked less than two months on the railroad when Bill was mauled by a panther and died. (At another time, John Bruce said Hatfield died of cancer. It is more likely, if he were killed by a wild animal, it would have been a lion.)

Alice Bruce and her two daughters were now in Durban as were Bill Hatfield's widow, John's sister and her child. John was heartbroken by Bill's sudden and tragic death. He also missed his wife, family and Aussie mates that were in Durban. He quit his job and set out for that port city. There he had the painful duty of telling Hatfield's widow what had happened. In this port city he now planned to stay. He had decided this before he boarded the train for South Africa's third largest city. He was looking forward to the reunion with his family and friends. He was looking forward to joining the Aussie colony there. Durban sits 250 feet above sea level. Its climate more to John's liking than Johannesburg. Its harbour was filled with ships of all nations. Its streets were crowded with wagons, shouting teamsters and carriages, buggies and horse-drawn cabs. It was a hive of industry, a busy metropolis. The depression that had brought Johannesburg to its knees had not yet effected Durban. John was born and raised in a seaport, so Durban seemed like home to him. He rented a flat from Robert Young, a sergeant on the local constabulary. Bruce and his landlord became very good friends.

John found a job at J. Lormer's, a union shop whose work force numbered only twelve. The President and Secretary of the Durban plumbers' union also worked there. He joined at once and remained an active member until he left South Africa. The labour movement was weak and had little public support in South Africa. Nor was it supportive of the Labour Party. John, the humanitarian, soon was to become an embarrassment to his union colleagues.

He and his wife immediately became a welcome addition to the Aussie community. John became a very close friend with William Ritchie, a fellow Aussie who was the Cockshutt Plough Company sales manager.

Doctor Billie Maloney who practiced medicine in the city and John also became great chums. Maloney, an active Labour Party supporter, soon recruited John Bruce. He and Dr. Maloney were great soap box orators. This was before radio, television and talking pictures. There was no sitting staring at a idiot box or walking with a radio plugged into your ears in those days. One had to make or find his or her own amusement. Public forums were popular and well attended. Public gatherings were the only method for politicians and evangelists to communicate directly with the public. People attended to be amused, to learn or to argue. Hecklers added zest to these meetings. A soap box orator who could not handle his hecklers was laughed off his platform. Both the speaker and his heckler had to project their voices to be heard. The casual listeners enjoyed the exchange between the two adversaries. Because of the warm weather in Durban and the buildings having no air conditioners, outside rallies were popular. These forums were open to all who wished to attend. Some soap box orators became quite popular and influential. On most Saturday nights, one could walk down West Street to where it met the Espanade at the Municipal Gardens and hear either Maloney or Bruce. They both praised the virtues of the Labour Party. Each had his own way of silencing his hecklers. John was prone to coaxing his critic into making a ridiculous statement. one so silly the audience would start laughing and shouting the heckler into silence. Maloney and Bruce were skillful in handling those who taunted them and soon they were speaking to friendly audiences. More and more listened to their gospel of socialism than came to be entertained.

These two Labour Party spokesmen did not pass up any chance to criticize the "powers that be", the British authorities. The local newspapers often carried reports of the speaker's remarks. Their assaults on public figures were reported in the news columns. Maloney ran as a Labour Party candidate in the City of Durban in 1904 and was elected to city council. John Bruce was one of his busiest and most devoted workers.

The police attended these open air meetings and watched from the sidelines. Some plainclothes men were busy taking notes. On occasion, the police strutted about trying to intimidate the speakers and their listeners. Most of the time the cops just watched what the authorities thought were these potentially dangerous men. Pencils would fly over the pages when Bruce mentioned the plight of the blacks. Any voice of dissent worried the British authorities.

After the end of the Boer War, the economic boom began its decline. In 1904, the authorities brought in the Chinese to work in the mines. John was assigned to work on the compounds that were to house the Chinese labourers. These structures had been army barracks near the docks.

Economic conditions in South Africa were worsening. Conditions became so bad that white men were working for the same wage as the blacks. There were make-work projects, men digging holes in the sand, then filling them up again. The government was giving these workers barely enough to live on. The grave situation became so tragic it forced the British and Australian Government to act. Both countries made arrangements for any of their citizens who wished to come back home free. John had by now made up his mind to leave the country but not to go back to Australia.

Half the white work force were without jobs and others had suffered a serious reduction in their wages. Employers began cutting wages, first by a quarter, then by forty and sixty percent. Most had no choice. Accept the cut or be fired. The weak unions and the calamitous state of the economy gave them no choice.

Since coming to South Africa John had been employed at "pretty good" wages. He was superintendent in charge of the plumbers, erecting the new municipal building in Durban. One day, his employer, Mr. Lormier, came to him and said, "I am cutting the men's wages."

"You can cut the men's wages and what they do about it is up to them, but you're not going to cut mine. If you do, I will quit."

"John you can't quit and leave me alone with this whole job." Bruce had costed and planned most of the job. He knew its every detail; knew the job better than his employer. His wages were not reduced.

A few weeks later, the scene was again replayed. John was aware that the workers on the job knew his wages had remained untouched. They let their boss know that they thought that was unfair. He realized Mr. Lormier was being pressured to cut his wages. Again the employer said he was going to cut John's wages. John answer was the same. His wages remained unchanged.

Almost a month passed when John's employer came to him and again declared he was going to cut John's wages. John's reply was, “I quit. I’m going to Canada.

“You can’t quit me, John. I can’t run this whole job myself.”

“Send your son around and I will teach him all he needs to know about running this job.” John was serious and his employer knew it. Young Lormier, the son. also a plumber, came round. In a few weeks John had taught him enough so he could manage the job.

From the very day he landed in Africa, John resented society’s brutal treatment of the natives. He continued his vocal support of the Matabeles after the murder of Sergeant Armstrong. The authorities had tolerated his call for better treatment of the blacks. The whites knew that if the blacks were paid more, they would be paid less. Few of the whites and none of the Afrikaans were prepared to share with the blacks. Bruce could talk about sharing and it would do no harm.

The pressure on John Bruce by the British authorities was becoming intolerable. Taking the side of the natives kept getting him in wrong but he was no threat to the powers that be. His support of the blacks did not make him welcome among the whites. This was an unpleasant surprise for John as he expected better of them. They resented him standing up for the natives, especially after the Matabelian uprising in which Police Sergeant Armstrong was killed. Armstrong was trying to maintain the peace in the land stolen from the Matabeles. Blaming the British Government for their brutal treatment of the Matabeles was one thing, but blaming it for the sad state of the economy was too much. With the economy in tatters, this advocate of socialism had first been a slight annoyance, but now had become a problem. The establishment undertook to make Bruce’s life as uncomfortable as possible. He thought that he had better get out of South Africa. John’s audience had been listening politely to his socialistic views. That was when they had good jobs, good pay and a good life.

Suddenly, their lives had changed drastically. They had no jobs or the more fortunate had much reduced wages; most were in shock. Something had gone wrong with the system. Capitalism had failed them. They wanted the wrongs put to right. The socialism preached by Bruce and Maloney could be the answer they sought. This change in public opinion gave the authorities much cause for concern. One could say they overreacted.

John Bruce, an amusing and harmless dissident, had become a preacher. One that the unemployed listened to, nodding their heads in agreement with the gospel he espoused. Socialism and communism were the same to those with wealth and those that represented the government of England. Bruce was to be silenced. Throw him in jail on some charge. The English Civil Service, the Colonial Office often flouted the law and so had no hesitation in issuing a warrant for John's arrest. John found out about their decision from his friend and landlord. For the moment he had a safe hide-out. He was living with Bob Young, a sergeant of the police force who was supposedly trying to find him.

Bruce always had an ace up his sleeve. He had a friend, William Ritchie, who thought John would make a great sales manager. He asked John to consider becoming his replacement. He arranged for Mr. Mitchell, who had come from the Brantford, Ontario office of the Cockshutt Plough Company, to meet with John Bruce. If John accepted the position he would go to Brantford for a time to learn the business. The Canadian government was flooding South Africa with brochures listing all the country's virtues. It suggested the streets were paved with gold. It didn't specify which ones. John now dreamed of going to Canada but had not mentioned this to his wife Alice.

One night John arrived home with a handful of pamphlets

about Canada. Placing them in the hands of his wife he said, "Let's go to Canada." John had always been paid top dollar and had looked after his money. He had a nice little nest egg set aside, more than enough to travel in comfort to Brantford to take up a job with Cockshutt's.

Mrs. Bruce was quick in her response," No, let's go back home." John was a little taken aback by her reply. He countered by saying, "Oh no! let's go some place else — let's go to Canada." As the discussion continued John saw he had to give ground. His final proposal which Mrs. Bruce agreed to was, " Let's go to Canada and from there we could work our way home. Perhaps go to China and down from there"

Bruce decided he and his family must leave the country without delay. John bid his good-byes to the Lormiers for he had decided to leave South Africa in a hurry. It was much preferable to spending time in the primitive conditions of a Durban jail. Secretly he had booked passage for himself, wife and three children on a Castle Line ship for England. Since he heard of a warrant being issued for his arrest he had remained safely hidden in the home of the Police Sergeant Young.

The Castle Line ship left Durban at twelve o'clock noon, on the dot. As the clock struck twelve, the horn gave a couple of blasts. The hawsers were dropped and the boat left the dock and its departure would stop for no man. This company rule proved to be most helpful for John Bruce, a man on the lam.

When John and his family arrived at the point of departure but not together. The police were at the dockside waiting. A possibility that John had made allowances for. There, standing waiting for him to appear were his police sergeant friend, Bob Young and his superior, Inspector Alexander. His friend, Young, had given him forewarning of his pending arrest and John had taken the necessary precautions. He told his wife to board the ship without him, Alice with the children walked up the gangplank reserved for the passengers. John worked his way around to hide among the crates and boxes being loaded at the hind gang plank. Waiting until the last possible moment to rush from his hiding place to jump aboard. He had put his trust in the Castle Line's determination to stick to its schedule come hell or high water. Not to stop even if a police man yelled "STOP THE SHIP".

As the clock struck twelve noon, the horn gave its two blasts. The hawsers were being thrown off, John Bruce left his hiding place rushed and jumped aboard as the freight gang plank was being moved away. Inspector Alexander saw John rushing to board the ship and yelled "stop that man." A command that the Castle Line captain would not obey. He was not prepared to ruin the Lines' reputation for punctuality.

As the ship moved away from the dock. Bruce was now safely out of reach. Sergeant Bob Young pulled at the Inspector's sleeve and said "Let him go. We're better rid of him". Inspector Alexander then abandoned his attempt to arrest John Bruce. No serious effort was made by the police of Durban to hunt John down. So it was in haste and without ceremony John Bruce left the sunny shores of South Africa. Getting out of Durban had been John's Number One Priority. It was now safe to relax and think of "Faraway Green Fields". This John did as the ship sailed towards the British Isles.

Bob Young's protection of John was an example of the strong bonds of friendship that Bruce built. His loyalty to his friends, his faith and philosophy were complete. There are many examples of the loyalty he established with those associated with him.

GLADSTONE, 84.

Born December 29th, 1809 died May 19th, 1898

He was active in British politics for near half a century and served as Prime Minister. He was much disliked by Queen Victoria. A supporter of Cecil Rhodes, land pirate and murderer. Always willing to send troops and sailors to help Jameson and Rhodes suppress the Blacks so to add to the wealth of Britain's elite.

From the Review of Reviews, October, 1882

MR. RHODES; THE NAPOLEON OF SOUTH AFRICA
From the Westminster Budget (London)

Cecil Rhodes. Land pirate and murderer. Born in Hertfordshire, England on July 5th, 1853; died Cape Town, South Africa on March 26th, 1902. Graduate of Balliol College, Oxford. Amassed a great fortune by driving the Matabeles off their land at gun- point. If they resisted his hired thugs under his orders killed tem. William Gladstone Government in Britain approved of all his illegal activities.

English historians conveniently forgot all Rhodes' nefarious brutal treatment of the Blacks. He used some of the wealth he stole from the Matabeles to set up the Rhodes Scholarship. The deed does not make him any less of a thief and murderer not make him any less a murderer and land thief.

Photo from Review of Reviews, October 1893

Chief Lobengula of the Matabele tribe and his favourite wife. The Matabeles were a part of the Zulu nation, the Prussians of Africa. Photograph
from the October, 1893 issue of the Review of Reviews

Chapter Nine - Canada Bound

In March, 1906, the Bruces left South Africa on a Castle liner, John, a poor sailor, had a stormy trip from Durban to Cape Town. John the poor sailor was greeted by a rough, turbulent sea, The ship would rise on a wave then dive into its trough. His stomach churned as much as the sea. The five hundred miles from Durban around the Cape of Good Hope were sick days for him.

But from there to London John never had a more glorious life on the ocean. Now relieved of the worry about sea sickness, he joined in the social functions. He was a fine singer and soon was leading his fellow passengers in songs, story telling and reciting. As usually happened, he was the centre of attention.

When the liner glided into a calm sea of ripples and sun, John, never a good sailor, had a most pleasant journey.

John sat in the warm south Atlantic sun a few days after leaving South Africa. The gentle sunny breezes and the gentle rolling of the ship lulled John into a much-appreciated sleep. When he lay awake, he gave thought to what future may lie ahead.

While alone his thoughts would drift back to his brief stay in South Africa. The mind of the white South African Christian was a conundrum to him. Why were they so eager to convert the Blacks to their Christian faith, yet not treat them as human beings? The Bible-carrying white would not share a church pew with a Christian Black. In fact, they would not allow a Black to enter their place of worship.

Why did they assure the Blacks that they would spend eternity in Heaven? If they did not let the Blacks into their churches, were they prepared to share their Heaven with them? Did his former

Presbyterians believe that there was a 'hereafter' for them and another for the Blacks? John thought if that were true there must be a different Heaven for each colour? John thought he was remiss as he had not asked his fellow whites the questions.

In the warmth of the sun and gentle rocking of the ship he slipped off to sleep. Thus he was left with no answer to his conundrum.

With two daughters, shortly to enter their teens he had little time to worry about the whites he had left behind in Durban. He learned quickly that his wife expected him to keep his two older daughters, Hazel and Alice, amused. He did his best with games and stories. He was not much help with the six-week-old third daughter, Amy, although he made awkward attempts to be helpful.

He would get involved in a conversation with one or more of his fellow passengers and his daughters, bored with standing waiting, would walk off. John would look up and they would be gone. He would rush off to find them making a hurried apology . This would happen a few times a day. John was not a good baby sitter at the best of times. His wife Alice found him as much of a hindrance as a help.

John had arranged a stop-over at the Canary Islands when purchasing his and his family's passage. The Canary Islands, a Spanish territory in the Atlantic Ocean, was a port-of-call for this Castle Line ship. The Bruce family disembarked to enjoy a holiday. Whether Alice was aware of the visit to these islands when she boarded ship, John never said. They stayed on the Island of Las Palmas and then on Tenerife. He never did admit that his stop here was designed to confuse the police, especially those who may be waiting for him at the London dock.

He stayed in London for about five weeks, longer than he had planned. Some of it was involuntary. He had booked passage on the maiden voyage of the Empress of Ireland. "The C.P.R. had juggled my tickets" he alleged. He invaded the London office of Canadian Pacific Railways demanding satisfactory compensation. His argumentative, hostile manner intimidated the branch manager. His discomfort and embarrassment amused, and in some ways pleased, his fellow C.P.R. employees. To pacify this noisy complainant, the manager paid for John and his family's stay in London. Here they remained until they provided passage on another boat.

While in London, John visited the parents of a little English chap that he had befriended in Durban. His London host was the superintendent for the contractor building Whitehall, the new war office building. He offered John a job on the big new block being built to replace several older, out-dated structures. The Secretary of the London union of plumbers wouldn't let him go to work. They had several members out of work; some had been without employment for several weeks. John described the episode in these words, " I can't let you go to work when I've got all those men sitting on the bench. Of course, I had no argument. I was a good trade unionist. I could not expect to be allowed to work with all those men sitting on the the bench. Some had been without work for a long time. Conditions weren't very good. I could have gone to work but I refused it." He said that he had no desire to work there. He didn't like the climate or the working conditions. John didn't say whether it was the cold reception by the London plumbers' union or the dampness or the cool temperature that he found objectionable. John, at this time was set on joining the staff of Cockshutt Plough in Brantford, Ontario.

Alice's mother's relatives came in from Cardiff, Wales to

spend some time with the Bruce family. Alice had never met these kinfolk. All enjoyed the social exchanges and got caught up on the family gossip. John remembered this family visit and his pleasure at meeting his wife's Welsh relatives.

Five weeks after their arrival in England, they sailed for Canada. On the boat coming over, John met a carpenter from Toronto, Charlie Bishop, a member of Local 27 and a socialist. "He talked the idea of going to Brantford right out my head. Told me that I was a fool and that I should stay with my trade. There was plenty of employment in Toronto. He said I could make a contribution to the labour party but only if I was in Toronto. Brantford he described as being little more than a rural village, the back woods, removed from the world of union and socialist activities."

When the ship docked at Quebec City his decision to make Toronto his home had been made. John found his luggage which had been destined for Brantford. He painted out the word 'Brantford'. Then, he painted 'Toronto, Ontario' on each piece of luggage.

Very soon after his arrival in Toronto, he traveled to Brantford. There he saw the Cockshutt people. If all his money had not been transferred to a bank in Brantford, he would not have made the journey. He went there only to have his funds transferred to a Toronto bank.

When he saw the conditions under which the Cockshutt employees were working and their pay scale of eight and ten cents an hour, he knew this company was one he didn't wish to have as his employer. These hard facts made him more pleased with his decision to stay in Toronto and work at his trade. He returned home in very good spirits.

Soon after his arrival in Toronto and before his trip to the

Cockshutt plant he visited the business agent of the Plumbers' union, Bill Storey. John described Storey as a fine man. He told John that he would face tremendous hostility because he sounded like an Englishman. On returning from Brantford John found accommodation for his family and began looking for employment. He was the father of three girls, had a charming wife and was thirty years old. He looked forward to his future in a new country. He made this city his home until he died. Alice and he never got to visit China.

Chapter Ten - The Immigrant

A warrant for his arrest was still active. John never did explain why he accepted the offer from Cockshutt farm implement company. If he was being groomed to become their sales manager he would have to square things in South Africa before his return. While on board ship, the persuasive Charley Bishop all too easily convinced John to stay at his trade and settle in Toronto.

When he boarded ship he was bound for Brantford. His willingness and quickness to change his destination may have been for several reasons. Apartheid was so deeply rooted, John saw no end to it. His campaign against it had been pointless and the warrant for his arrest in itself was reason enough not to return There may have been other reasons but the two above were paramount.

This may explain him agreeing so readily to settle in Toronto. He may never have seriously entertained any thought of returning to Durban. He never indicated in any way that he was going back. The trip to Canada may have been enough to stop the 'itch in his feet'.

When he decided to accept the Cockshutt job, he may have thought that he had no alternative. Charley Bishop gave him a more attractive one. John, who almost invariably engineered it so he had two choices, appears to have had only one until he met Charley Bishop.

This is John Bruce's version of his flight from South Africa and his decision to settle in Toronto, the city in which he was to reside for the rest of his life.

By the time of his arrival in Canada, Charley Bishop gave John a complete rundown of the union and political situation. Part of the information he acquired was wages paid, working conditions, union politics and strength, and who was in charge. Much of his success in his lifetime was his ability to obtain advance knowledge about the people and the problems he was to face. Thus he was rarely caught unawares.

Just under a quarter of a million people lived in Toronto when John arrived. The City Council had by then passed a by-law permitting the trolley cars to run on Sunday. Barber shop quartets had just added 'Sweet Adeline" and "The Good Old Summer Time" to their repertoires.

The city he came to had a serious fire in 1904. A roaring blaze destroyed fourteen acres in the city centre. It began in E. & S. Currie Neckwear Company on the north side of Wellington Street and spread rapidly. By the time the blaze was extinguished, one hundred and twenty structures, including over eighty commercial buildings, were only piles of smoldering ashes. The resulting re-building of the downtown area gave work and high wages to the men in the building trades. Their wage rates were approximately four times more than those of factory workers. All the new structures, unlike the old, had built in all the modern plumbing fixtures of that era. Iron pipe and fittings, drains and sewer installation were now required by the sanitary codes then being introduced by local governments. The revised building by-laws provided added work for the plumbers. When Bruce arrived in 1906, the building boom was entering its final phase. Building trades men were still much in demand.

Before going to Brantford, he settled Alice and the three children in the hotel at the corner of Simcoe and Wellington Streets. It was more of a boarding house than a regular hotel. It was located near the railway station and a little west of the city centre.

Bill Storey, Business Agent for Local 46 of the United Association first sent John to a plumbing contractor whose yard was just east of Church Street. John walked through the yard to a building whose paint was so weathered it was difficult to determine its original colour. He opened the door without knocking and entered a smoke-filled room. Sitting in an oak swivel chair was a small man wearing a three-piece suit of heavy tweed. A well- seasoned briar pipe was clenched in his teeth.

John bid him "Good morning."

His greeting was answered with, "We don't want any bloody Englishmen."

The Scots, Welsh and Irish had long suffered the mal-treatment afforded them by the English ruling class. Hatred of the English was not just reserved for the rich. The Irish, Welsh and the Scots had enough hatred so they could lavish it upon all the English. Even those as badly exploited by the English upper classes as were the other ethnic groups living in the British Isles. These three non-English groups in Australia had the same hatred of the English. In the land down-under, the English were referred to as 'Palmy Bastards'.

John likely would not have reacted so vigorously had his greeting come from someone other than an Englishman. The man who said, "We don't want any bloody Englishmen." had a much ingrained English accent.

John was so taken back by this response he paused before phrasing his reply. His prospective employer, thinking John did not hear him, said, "Did you hear what I said? We, don't want any bloody Englishmen."

"How are you to judge, how do you know I'm an Englishman? You said it, but how are you making out I'm an Englishman?"

John's prospective employer was a Yorkshire man and his upcountry brogue could be cut with a knife. John, who was four to five inches taller, squared his shoulders and a long harangue followed. During the exchange of mild insults, the Yorkshire man realized he had make a mistake. Exactly what it was he was not sure.

He backed off a bit and asked "Where do you come from?"

"It doesn't matter where a man comes from or his nationality or the colour of his skin if he's a good mechanic."

"I never met any Englishman that was any good."

"I've come to this country from a long way to find a man of your standing."

John's prospective employer had no answer. He said, "However I'll give you a chance, you come to work Monday."

"You've got the wrong man. I wouldn't work for you if I was starving to death. You don't know how to treat men humanely. I'm not going to work for you because I couldn't give you service."

He accepted John's refusal with some grace for he said, " All right, I'll send you over to another shop." John didn't go to the other shop mentioned. Instead he went back to the plumbers' union business agent.

The U. A. Local 46's Business Agent Bill Storey's office was in the Toronto Labour Temple in 1906. The man like John wore a three piece suit, a derby and sported a suitable moustache. When John told Bill Storey that he refused to accept the job offered he was told to go where the employer suggested.

John followed Bill Storey's instructions. He entered the office of Bud Long to be greeted by two Yorkshire men. Again he was told, "We don't hire any Englishmen."

John looked at him and asked, "Well, when did you renounce your country? It would take a rough file to get that Yorkshire burr off your tongue. I don't happen to be an Englishman, not that that is any disgrace. I am surprised that an Englishman would stoop to that level."

They looked at each other and both laughed. Then one said, "We were told you are quite a tiger and caused quite a stir."

"This is no laughing matter." said the irate Bruce.

The amused Yorkshire men told John that they had no work for him but that he should go to another plumbing shop. There he obtained employment. The company was Fitzsimmons and Company where John worked from June, 1906 until the 1907 strike.

Chapter Eleven - The New Member

When John Bruce completed his apprenticeship he had joined the Melbourne Branch of the United Operative Plumbers' and Domestic Engineers' Association of Great Britain and Ireland. He transferred his membership to the Durban Branch when he settled in South Africa. In the spring of 1906 he arrived in England and within a few days went to the United Operative Plumbers and Domestic Engineers' union at 15 Abbeville Road, Clapham, London S.W. 19. All branches accepted his transfer card without question.

As was his practice, John joined the local union at the first opportunity. His membership in the English based union in Durban was no help to him here. He had to appear before the local union executive and if accepted as a member he would have to pay a seventy-five dollar initiation fee. This did not sit too well with John Bruce. He knew he had no choice.

However soon after John arrived in Toronto he walked from his hotel on Simcoe Street to the Labour Temple at 167 Church Street. He went to the office of Business Agent Bill Storey and presented his transfer card. Storey told him the transfer card was not acceptable. The United Association had no reciprocal agreement with its British counter-part. Little wonder. Who would give up a job with better working conditions and higher wages to go to class-structure ridden England or Ireland?

At first John protested and reminded Storey that all union men were brothers, kindred souls. The business agent raised his hands above his head in exasperation and announced "That's the decision of the head office and I cannot change it." He opened his desk drawer, took out a copy of the United Association's constitution and told John to read it. This he did. He read it so carefully and thoroughly that he became Local 46's constitutional expert. He used this to his advantage.

Storey told John that the initiation fee was seventy-five dollars, a large sum in 1906. John accepted the news without protest which pleasantly surprised the business agent. For a plumber earning thirty-eight to forty-two cents an hour it was a very large expenditure. This fee was rarely paid in a lump sum. Most paid an extra dollar or so when they paid the thirty cents a week union dues until the debt was paid off.

Shown below is the application card that John signed in June of 1906.

APPLICATION FOR MEMBERSHIP

To the Secretary __________________________ Journeymen
{City or Town}
Plumbers' Protective and Benevolent Association:

The undersigned represents that he is a Competent ______________________ plumber and desires ad-mission to membership in your Association; and I further agree to live up to all its laws from time to time as amended.

The Association reserves the right to reject any application they see fit.

My name is..

My age is ....

My address is..

Have you attended any trade school? ....

Where ....

How long have you been at trade? ....

Where was your apprenticeship served ? ....

How long did you serve as an apprentice ....

How long did you serve a Junior ? ....

How long did you work as a Journeyman ? ....

Did you ever make application to the U. A..

Were you ever initiated by any Local of the U. A. before?

....

....

{Name Local and town where initiated.}

Did you ever owe any back dues to any local of the U. A.?

....

{Name town, city or state}

Were you ever fined by any Local of the U.A. ?

....

{Name town, city or state}

Have you ever worked during a strike or lockout ?

....

{If so, name place}.

Where last employed ?..

Where employed at present ? ....

Did you ever work at trade under any other name ?

....

{If so, where?}

Enclosed please find .... for application fee,

I further agree to pay balance at 25 per cent per week; and it is understood that nothing less than 25 per cent per week of wages earned will be accepted as part payment of initiation, and that if the total initiation fee of seventy five dollars be not paid in installments of not less than 25 per cent within ten weeks from the date of this application then all action hereunder to be in the discretion of the Examining Board, to be declared null and void, and all moneys paid upon the same by the candidate to be forfeited by him to the Association as compensation for the time consumed, expense incurred and trouble endured in the matter of the candidate's application. I also agree to forfeit to the Association the sum of five dollars per week, to be retained out of my deposit for privileges enjoyed during the pendency of this application. In the event of Examining Board rejecting my application.

Signed ....

The Examining Board has investigated above applicant and finds him

.... to membership.

Chairman of Examining Board ....

Member of Examining Board ....

Member of Examining Board ....

Application Blank

We, the undersigned members in good standing of the

....Journeymen Plumbers

{city or town}

Protective and Benevolent Association, do respectfully recommend Mr.

....

{space for Gas, Steam or Sprinkler Fitter}

as a competent Journeyman Plumber, for membership in the Association, providing the applicant's statements are true, with his signature attached

....

Caution to Members

Any member vouching for a person who is not a competent journeyman plumber shall be deprived of such monetary benefits as he otherwise would entitled to for the period of three months following.

The questions asked on the application card give one the impression that the United Association had a serious problem with delinquent members attempting to sneak back in by joining as new members to avoid paying back dues. Yet, with an initiation fee of seventy-five dollars it is unlikely that anyone owing less than five years back dues would be prepared to pay the initiation fee a second time. It appears the U. A. was out to catch errant members who may have worked during a strike, failed to pay a fine levied or skipped out owing back dues.

Like most craft unions the initiation fee of the United Association was very costly. While the dues in 1906 were only thirty cents a week, the initiation fee was equivalent to almost five year's dues. Hourly wage rates of plumbers in 1906 were between thirty and forty-two cents an hour. A plumber or steam fitter making forty cents an hour would have to work one hundred and eighty-seven and a half hours to earn enough to pay the seventy-five dollar initiation fee. With the advent of the Committee of Industrial Organization {the CIO} initiation was reduced to the minimum required by the labour laws.

In the early 1900's, each plumber applying for membership in Local 46 had to appear before the Executive Board. The board members wore blue serge, three-piece suits, some grown so old that the dark blue had a brassy-green tint to it. It was the conventional wisdom of the day, that a man was married and buried in the same suit, by which time it was a pronounced brassy green. Each Board member wore a white shirt, some not so white. Each shirt had a detachable celluloid collar. These collars were not washed but cleaned with an Indian rubber eraser. Around an oak, ink-stained table they sat. Most smoked. Some puffed away on their briar pipes while others smoked inexpensive locally-rolled cigars. The small office, with its high ceiling, was smoke filled. The other furnishings in the office of Local 46 were ten wooden chairs; all placed around the oak table that occupied the centre and most of the room. Against one smoke-coloured wall stood the business agent's roll-top oak desk. Its pigeon-holes were stuffed with envelopes, pamphlets, papers and circulars. Almost within reach of it stood two or three-drawer oak filing cabinets. On top of the aged desk sat two wooden index card boxes which held the membership/ledger cards of the members.

At one end of the table, sat Business Agent Storey. At the

opposite end sat the applicant for membership. This particular night, John Bruce sat there, better and more fashionably dressed than the others. On each side of the table with its several ash trays, sat the members of the Executive Board, all looking wise and official. They put their questions to the applicant for membership. Some questions were irrelevant but they followed a time worn pattern. "Where had he served his apprenticeship and with what contractor?" "Where had he worked before and where did he work now?" After many questions had been put to the applicant he was asked, "What hourly rate are you paid." All inquiries John Bruce answered with ease and composure, save the last. When asked what his wage was, John hesitated, he appeared reluctantly to answer. Then, he said, "Forty-two cents an hour."

The Executive Board member sitting one man away from John shouted, "That's a lie. No new employee at Fitzsimmons gets paid that much." John with much quickness stood up, reached around the man next to him and grabbed his accuser by his shirt front. He arched his other hand, now a fist, to smash his potential victim squarely in the face. The man, now crowded between the two antagonists, grabbed hold of Bruce's arm, thus preventing the blow being struck. After some shuffling and righting of a couple of chairs that had been knocked over, peace or rather quiet was restored.

It was then that John Bruce reached into his inside coat pocket and took out his 'pay packet'. A small manila-coloured paper envelope on which was neatly written the name of its owner, his rate of pay per hour and the number of hours he worked in that week. On this envelope was John's name, the hourly rate of forty-two cents and the forty-four hours he had worked the previous week.

His accuser, a fellow who worked for J. Fitzsimmonds and

Company now fully recovered from his few moments of danger, was full of apologies. John's application for membership in Local 46 was quickly approved by all.

John Bruce's first two attempts to obtain employment had provided some amusement in the community of plumbers. The news of this traveled fast and was now common knowledge. The Executive of Local 46, like most, had heard of the recent exploits of this renegade immigrant plumber. No doubt this made an impression on those who approved his application. Like a new kid on the block, all eyes were now focused on John. His feisty, aggressive and imposing manner, loud voice, size and stance marked him as a person to be reckoned with. Little did the Executive Board realize the impact this new member would have on the local, on the United Association and the labour movement generally. John was ready to play his part in the local, national and international labour movement. The author is not certain whether these groups were ready for John Bruce.

The morning after this hearing, John delayed beginning his workday and paid a visit to the office of his employer. Somewhat nonplused, he apologized for breaking his commitment not to divulge his hourly wage rate. When the contractor agreed to pay John the forty-two cents an hour, he had agreed not to reveal his rate of pay. He explained the situation that forced him to break his pledge. His boss nodded his head indicating he understood and said, "they would have found out sooner or later, no secret can remain a secret long in this trade." Having squared accounts with his employer. John arrived at the work site an hour late.

John Bruce, because of his rough confrontation with his first potential employer, had gained some notoriety. His altercation with a member of the Executive board of Local 46 gained him more attention. He was becoming known as a man with a witty, bitter tongue and a short fuse.

Once his application for membership was approved he could attend membership meetings of the local union. He did but it only added to his reputation as a scrapper and a determined debater.

The regular meetings of Local 46 were held in Room One of the Labour Temple, 167 Church Street, Toronto. In 1904 the building was purchased from the Athenaeum Club, a 'gentlemen's' club (or one that pretended to be} . The Toronto Trades and Labour Council set up the Toronto Labour Temple Company to make the purchase. In those days in Canada, unions were not considered to be legal entities and could not hold title to property.

Room One and Room Two of the Temple were almost identical. The only noticeable difference was the ornately carved wooden throne chair used by the chairman in Room Two. These two rooms which had served as the dining room and lounge of the Athenaeum Club, were long and narrow, measuring twenty feet by seventy feet. In one end of these rooms, with the sand-coloured painted walls protected by layers of tobacco smoke, were two windows. They let in little light and were rarely, if ever, opened. In the morning one could not avoid the smell of stale tobacco smoke. By mid-evening, the room smelled of a mixture of stale and fresh pipe and cigar smoke. Truly, it was the age of the smoke-filled room.

In Room One, the chairman occupied a high-backed oak armchair throne chair which sat on a low platform. He sat behind a small oak but ornate table with a fifteen inch square top; more a stool than a table. The Secretary and Treasurer of the local union sat at a small table at floor level. These two tables were midway along the seventy foot long wall. Facing them were four rows of folding wooden chairs, that came joined in sets of four. When the members sitting towards the extreme ends of the room rose to speak, they half turned to face the officers in the room's centre.

The auditorium could comfortably seat about three hundred and fifty to four hundred. It had been the gymnasium of the Athenaeum Club. It had a narrow balcony that completely encircled the room. The balcony was used, on occasion, to hold extra persons. They had to hang over the railing to see what went on down below. The balcony had been a track around which members of the Athenaeum Club ran to keep fit. Because of its high ceiling, tobacco smoke was less of a problem.

The auditorium was the room where the mass meetings John Bruce referred to were held. Special meetings of Local 46 of the U.A. to take strike votes, or to ratify new contracts, were held in this room. When John joined the Toronto plumbers' local it had a membership of about five hundred. The auditorium was an ideal meeting place for such gatherings.

The union members arrived at these meetings wearing the usual blue suits with white shirts. Neatly knotted narrow ties were in fashion. John Bruce was, with few exceptions, the best and most fashionably dressed member of Local 46. He always took great care and pride in his appearance.

When he attended his first membership meeting he was in for a surprise. To his amazement he found the anti-English resentment prevalent. A little Englishman, named Teddy Britain, got up and in a real broad Lancastershire accent said, "Mr. President, I should like to know——"

He was interrupted by someone yelling, "Sit down you god damned Bronco."

John was on his feet with lightning speed and said in a loud angry voice, "Mr. President is this a trade union or some religious society, or am I in the wrong place?"

Bruce was told to sit down and his reply was. "I won't sit down until you make that man withdraw the remarks or him and me will go outside and settle it."

An old Irishman, named Arthur Lowrie, sitting near Bruce, said. "You sit down you god damned Bronco."

"Brother, it is lucky for you that you are the age you are, because I'd wipe the floor with you."

John said a half century later said that his fighting words "brought the house down. When I was going out to fight, I was accused of disturbing the meeting." It was all settled when someone tipped off the member who yelled the original order to John to sit down. Bruce wasn't English but established his reputation as being quick-tempered and a man prepared to settle disputes with fisticuffs. A mixture of apologies and a call for order, quickly brought order and decorum was restored. This episode established John Bruce as a member to be reckoned with. This and his other interventions made him, almost overnight, a very popular member of Local #46 of the Plumbers union.

Shortly after this episode, he was nominated as Treasurer of the local union. He accepted and he was elected.

In the meantime, he had become acquainted with Jimmy Simpson, Alex Lyons and Davie Crombie who were very progressive fellows around the trade union movement. All of them were declared Socialists. John threw his lot in with them as there was no Labour Party and no evidence of it.

Decades later John's comment was, "Strange as it may seem,

most of the trade union leaders of that day were identified with the old political parties." In 1907, John was elected as a delegate to the Toronto Trades and Labour Council. At the first meeting he attended he was surprised to see the members of council lined up; "the Conservatives on one side and the Liberals on the other. and a few who sat at the end were of no political faith. Religion played an important part in the labour movement. Many of those on the Liberal side of the house were Roman Catholics. On the Conservative side many of them were Orangemen. I never before knew religious antagonism in the union movement. I had heard of Orangemen in Northern Ireland and Roman Catholics fighting, beatings, and so on but never expected to find them here (in Toronto) I found the feelings to be very intensive."

Chapter Twelve - The Strike Leader

At 8.04 on the night of 1904 a fire started in the premises of the E & S Currie Neckwear Company, Wellington Street. The fire quickly spread. The low pressure in the water system made containing the fire most difficult. The fire raged throughout the night. Fire fighters from nearby cities, such as Hamilton, brought their equipment on flat cars to help fight the blaze. When the fire was finally put out, fourteen acres were devastated. One hundred and twenty odd buildings were destroyed, eighty-six were large commercial buildings. Then began a great building boom. Downtown Toronto was changed completely. Torontonians could hardly believe it; the change was so great. The building boom that followed the 1904 fire came to an end. There was less demand for building trades workers, including plumbers. Plumbers' wages had topped out at forty -two and a half cents. The master plumbers decided to use this down-turn to hold rates down,

The master plumbers saw the building boom coming to an end. Time to roll back wages, they reasoned. Encouraged by the Canadian Manufacturers Association to fight unions, they decided to provoke a strike. They, or the majority of them, believed the time was ripe. The economy was slowing down, unemployment was on the rise, so they thought they could afford a work stoppage. The employers did not anticipate such a total and clever resistance from the plumbers of Toronto. Employers deliberately provoking a union to strike was not new in 1906. It is still a practice used by some employers to take advantage of a depression or recession. In 1907 negotiations in the plumbing trade was a sham. The master plumbers got the strike they wanted.

John Bruce reported: “In May 7th of 1907, we {Local Union 46} came out on strike. We were unable to come to an agreement with our employers. They elected me to be on the Negotiating Committee; there were five of us. We were unable to reach an agreement. The employers {the Master Plumbers Association} were smarting under the fact that they had been fined ten thousand dollars for violating the Combines Act. In 1906, when I came here, there were only three or four non-union shops in the town. The employers had been forced to become members of the association or be ostracized. The union was exonerated. They couldn’t find anything in our books that tied us in with the combine. On the employers side, they found many instances of collusion to set prices. Therefore they were fined ten thousand dollars.”

“In the meetings to work toward agreement they didn’t want to give us anything. We were looking for a five cent an hour increase. {which would have set the plumbers’ hourly wage at forty-two to forty-seven and a half cents.} They wouldn’t give us anything, so on May the 7th, 1907 we went out on strike. I was put on the strike committee.” John with his evangelical eloquence, no doubt, was the most effective advocate of strike action. Little wonder he was put on the strike committee. “As Treasurer of the local, of course, it was a good spot to have me as well as on the Strike committee.”

On April 13th, 1907, twenty-five plumbers struck against Messrs. Martinter and Company. These were members of the International Association of Steam Fitters. This was the much hated rival of Local 46 of the United Association of Plumbers, Gas and Steam Fitters.

On May 7th, 1907, 500 members of Local No. 46 of the United Association struck against 104 contractors. Their demand was for an increase and the employment of union labour only; the 'closed shop'.

The Labour Gazette of August, 1907 reported *"No settlement was reported with regard to a strike of plumbers at Toronto which took place April 3rd affecting about 50 men on account of the employment of non-unionists and a general strike involving 500 plumbers and 104 firms, It was reported that at the end of July about half the strikers had left Toronto to seek work elsewhere and about 250 men being still out of work at the close of the month."*

"During this one-year strike some interesting events took place. We were making considerable gains. It looked as if they would have to surrender. Immigration was at its peak at that particular time and the Canadian Manufacturers Association was supporting the employers. A fellow named Jim Morris who was Secretary of the Association was very successful in hurting the strike at every chance he got."

"I think we were at a turning point, when the employers pulled one of the foulest moves I have ever seen. I am surprised to this day that some of our men fell for it. They picked about fifteen men in the five leading shops and gave each of them one share in the business. The men believed that as shareholders they could no longer belong to the union, and therefore were obligated to go to work. Most went to work, although four or five refused. That gave us an awful jolt. The plumbing shops of Fred Armstrong, Louis Leblanc and Martin and Johnson wouldn't go along with it. They thought this was a dirty trick; but it worked."

The September, 1907, issue of the Labour Gazette reported the following, *"The two strikes of plumbers at Toronto, one which began on April 13th on account of the employment of non-unionists and the other on May 17, "Their demand was for an increase and the employment of union labour only, the 'closed shop continued throughout the month. A strike of about 25 electricians employed by 5 firms who went out in sympathy with the plumbers continued throughout August. Messrs. Keith and Fitzsimmons one the firms said the Electrical workers violated an agreement. The union (the electricians) said that there was nothing in agreement as they had never forfeited their right to call a sympathy strike."*

"We found out what was going to take place through our usual sources, some of the employees who had been members of Local 46, and were still loyal. The men who were returning to work met on Victoria Street one night. Our Strike Committee approached these men and it almost turned to fisticuffs. They were determined to go to work. We were very disappointed."

"For a couple of weeks, the loyalty of our members couldn't be questioned but to see so many of our leading members of the trade going to work and being considered part of the firm simply by holding one share, gave them the feeling that all was lost. However, I was a kind of crusader and was able to rally their spirits during the strike meetings so they became enthusiastic to win. They were good guys and very loyal." Anyone who heard John speak could easily understand how he could rally the troops; restore their dedication and spirits. "It was surprising the number who stayed loyal," John said. "Five hundred had started the strike. We had a number of jobs with fair contractors, contractors who had signed for our rates and were working under our conditions and those men working were paying an assessment* to keep the others going." *{This assessment provided funds to assist those still on strike.}

"There was a surplus of labour due to to immigration of the lesser skills" reported the Federal Department of Labour's 'Labour

Gazette'.

"Trade conditions and the economy started to slacken then. We had the support of some of the trades. The bricklayers didn't support us very much, although they were a very good organization. The Labourers and the Plasterers supported us. The Electricians came out with us on strike at the same time. The Electricians were well organized and got a settlement before we did."

"Some of the employers tried in a number of ways to affect the settlement. Those employers, along with the Canadian Manufacturers Association, were out to give the strikers a trouncing. It surprised all of us the extent they would go to use the immigration laws of that time to recruit scabs. The employers and the C.M.A were advertising in England for plumbers. Those accepting a job could have their boat and train fare paid to go to Toronto. The fare at that time was very small, about twelve to fourteen dollars for a boat trip from Britain to Canada."

"We were appalled at the number of men that were coming from Britain. They had news of good conditions, you see. We immediately got in touch with the British trade unions, especially the plumbers union in Great Britain and had them advise their local unions, not to respond to any of the advertisements of the Canadian Manufacturers' Association. In addition we put a lot of advertisements in the union papers. We re-imbursed them. The local had fairly good funds at that time."

"Then, we found that many who were coming had union cards and we couldn't understand that. So we thought we would try a game on them. One of our members, Tommy Willis, a good union fellow from London {England} was a very active member of our union. We arranged for him to go to Montreal and meet the ships coming in and intercede with newly arrived immigrant plumbers. His job was to persuade them not to go to Toronto but to other parts of Canada. Tommy went to Montreal and was able to ingratiate himself with some of the immigration officers. Before the ship docked he had learned who were the plumbers and those who were coming to Toronto. Because of the immigration policy and the conditions that existed, the railway fare from Montreal to Winnipeg was ten dollars and one quarter of a cent for each mile after that."

"Of course, our west was growing tremendously at that time and there was quite a demand for men in Winnipeg, Calgary, Edmonton, Brandon and Swift Current, and Regina — all those western towns. The employers couldn't understand it. They would get word that a certain number of plumbers were on the boat coming. They found out we had somebody undercutting them in Montreal and were sending them west. They screamed."

"When winter came, immigration officials went to Saint John and Halifax. We sent Tommy Willis there. The photo engravers in Toronto were on strike at the same time and adopted the same principle. They put a man on and they were able to win their dispute by sending the men to other places. Photo engraving at that time was very limited in scope— there were very few mechanics available. Because of this they were able to win their strike."

"Local 46 persuaded some of their members to take advantage of low fares and go west to find employment. This reduced the cost of strike pay. Some of the plumbers who took up the offer became plumbing contractors. Although most are now dead (1969) some of the firms still exist. There was Nepiss, Austin & Jack Rogers all in Calgary; Fitzgerald in Edmonton; Holland and Grakawber and Beaton in Saskatoon and Joselyn MacLeod in

Winnipeg."

The October, 1907 Labour Gazette, *"A strike of plumbers which began on May 17th continued throughout the month. According to a report, received from the Employers Ass'n. the cause of the dispute was the demand on the part of the men for recognition of their union, a minimum wage of 45 cents an hour, limitation of apprentices, the requirement that apprentices should learn their trade in the City of Toronto and the restriction of use of tools with regard to apprentices."*

"We stayed on strike. We reduced our strike committee to five members. There were fifteen men when we started. Two or three of them went into business, one of them became a salesman and a couple of them went to work for fair contractors. The five of us decided to keep the strike going." It's possible the other four on the committee, like John, couldn't tolerate losing.

"We kept the strike going for a year. When some of the independent contractors broke away from the general contractors, they formed an independent group of contractors and signed with us after a year. The wage settlement was forty-two and half cents an hour."

"I didn't go back to work until I got forty-five cents. I wouldn't go to work and stayed out a further two weeks until I got my forty-five cents. The boss I had been working for was one of the members, Tommy Tyler, who had gone into business. He came after me and I told him that I wouldn't go to work for forty-two and half cents an hour. I had been holding out for forty—five and still was." Tommy Tyler agreed to give John his forty-five cent rate.

John said, "The strike ended, but we lost quite a number of our members, an awful lot of our members."

"We got strike benefits of five dollars a week for fifteen

weeks from the General Office {of the U.A.} and we had funds in the local union. Those who were working were paying twenty-five percent of their wages. Even in those days, they were willing to do it. We were able to pay strike pay throughout the whole thing. I will admit I had tough limits." John always had a nest egg but in October of 1907 he had none. He was forced to move from Boswell Street to the 'slums', Reid Street, where the rent was ten dollars a month.

In 1907, bread was five cents a loaf and milk was five cents a quart. One ton of coal cost four dollars and fifty cents. Meat cost six to twelve cents a pound. The average building tradesman was paid twelve dollars a week; labourers about eight dollars and plumbers were paid twice as much.

When John returned to work, he and his family moved to a semi-detached house at 154 Spruce Street. The rent for it was fifteen dollars a month. In 1969, when John told his life story, that house was still standing.

John does not mention the trials, anguish and difficulties his wife, Alice, had in surviving for a year on twenty dollars a month, a little more than John made a week when working. As in all long strikes the ordeal of the wife trying to look after her family is little noted.

More than sixty years later John made these comments about the events and the time: "The effect of the Plumbers' strike on the building trades was particularly 'destructive' because the Bricklayers, while they gave us support on quite a few jobs, did not do much in the way of helping us (the Plumbers). They were the premier organization in Toronto at that time in the building industry."

"Although, one admires the strength and youthfulness of the Labourers' Union, it was made up principally of old country people who came to Canada and followed the ordinary tradition of the building industry. Of course, everything then was hand labour — the mixing of mortar and the carrying of it up a ladder in a hog. The Labourers had one of the best organizations in Toronto. In fact, they had grown so big and were so well- established themselves, they bought their own hall. They bought the old St. Andrew's Hall on Simcoe Street, lower Simcoe Street. They maintained a couple of business agents at that time. They were very strong. In fact, they could tie up the Bricklayers anytime they wanted to."

"The rest of the trades, as the Electricians, for example suffered and the Carpenters, in my opinion, had the worst competition of all the building trades. A large number of non-union carpenters were in town. They found it very difficult to organize. There were two groups of carpenters. There was the United Brotherhood of Carpenters and there was the Amalgamated Carpenters of Great Britain. Those who were members of the Amalgamated Carpenters were very aggressive, good tradesmen and good trade unionists.

Of course, the same could have been said of those in the United Brotherhood but the town was so overwhelmed with carpenters. The economy was progressing pretty well at that particular time (1907). The fight for membership between those two unions, of course. kept that division wide open. There was no real organizing effort being made. Jack Garbott was the business agent of the Amalgamated and Albert Saunders was with the United Brotherhood, a very faithful, splendid fellow; I admired him very much. He was one of the directors of the Labour Temple and gave an awful lot of service."

"We were making efforts to re-build the building trades. At the end of 1908, I was elected President of the Building Trades Council. We started an organizing campaign in 1909 and we visited every local union in the building trades in an effort to stimulate some activity. The results were surprising. We visited various parts of Toronto, holding open meetings, either on street corners or in some of the halls we could secure. There were very few halls available. We sometimes had to use the back rooms of some of the pubs." (This must have irked John, the temperance advocate.)

The tightening of money supply threw the country into serious recession that frustrated the organizing efforts of John Bruce and his organizing committee. Their intention to strengthen the building trades failed. John summarized the situation thus, "We (Local 46) didn't get back on our feet until the war started in September, 1914."

Local Union No 46 lost almost half of its members during and after the 1907-1908 strike. Most were expelled for the non-payment of dues. The cost of again joining the local union was a price few thought they could afford.. An expense that they could not pay or would not pay. The solution, as the Executive Board and membership saw it, was to get special dispensation for their delinquent members. To do so the Local Union No. 46 submitted a resolution to the 1910 Convention of the United Association. (See below)

This is the verbatim report from the Monday afternoon session, September 26th, 1910 of the proceedings of the United Association of Journeymen Plumbers, Gas Fitters, Steam Fitters and Steam Fitters' Helpers of the United States and Canada, Convened in St. Paul, Minneapolis. {Page 129 & 130}

Resolution No. 158

By Local Union No. 46

Whereas, This Local Union has found it necessary, owing to the large number of members expelled for non-payment of dues subsequent to our year's strike in 1907 and 1908 to reinstate

a considerable number at reduced rates; therefore be it

Resolved, That all members so admitted shall be placed in the same position on the books of the U. A. as though they had paid the full reinstatement fee.

The committee reported unfavorably on the resolution.

A motion was made and seconded that the recommendation of the committee be concurred in.

Delegate Bruce - Do I understand by the recommendation of the committee that those men who were granted special dispensation after the strike of 1907 will still have to pay the dollar stamp to the head office or it will be held against them there?

Chairman Clark —The understanding of the committee is that where necessary for dispensation to be issued requests are made on the General Executive Board for such dispensation, and the men on the payment of the amount in compliance with the dispensation issued are in the same standing with the men paying regular initiation.

Delegate Bruce — But the general office is holding men for reinstatement stamps today who have been reinstated on the five dollar proposition. Probably if Secretary Burke were here he would know one case in particular. One of our brothers was reinstated on the five dollar proposition. Ten or twelve months after he made application to the general office for benefits for him. We found the general office was holding him for reinstatement stamps. If Brother Peller is held for that reinstatement stamp there will be sixty or seventy in the same position,

Delegate Groeniger — I move as an amendment that the subject matter of the resolution be referred back to the committee **(Seconded)**

Delegate Koch — I don't think that is necessary. We considered the matter. We had Brother Bruce there and he explained the case to us. We came to the conclusion that the proper thing for him to do is to bring this matter before the Grievance Committee or the Executive Board.

Delegate Bruce — I am not objecting to the report of the committee. I understood that the subject matter would be referred to the Grievance Committee.

Delegate Herderson — I move as a substitute that the subject matter be referred to the incoming Executive Board. {Seconded and carried}

John was now aware of the fact that the table officers were in control of the convention. His only hope of getting his resolution passed was trust them to do the political expedient. In which case all parties saved face.

Chapter Thirteen - The Good Samaritan

"The depression set in late in 1908 things started to get real bad. Immigration started to let up a little. There seemed to be a tightening of money at that time. A couple of our banks failed. The Farmers' Bank and the Home Bank had failed. York Trust, which was carrying on quite a development just west of Roncesvalles Avenue, they had also failed. They stopped almost dead in their building operations and threw a lot of people out of employment."

There was very little activity going on anywhere. Work opportunities were very limited around Toronto. "In the building trades we went to the city hall and met with the Board of Control urging them to start some public works. One of our members, J. J. Ward, a tailor, was on the Board of Control. Even, though he owned his own shop he was still a member of the Tailors' Union. He developed the idea of building a sea wall. Others laid claim to the idea but it was J. J. Ward who was the father of the idea. That's the wall that now runs from the Humber River to the Exhibition Grounds. Its immediate purpose was to provide work for the unemployed. The city fathers approved of the sea wall project. However, as kind of an unemployment proposition, they started to develop Sunnyside Beach. This became a place to sun or swim and picnic. They built the sea wall by putting pumps out in the lake and pumped in the sand from the lake bottom. Prior to building the sea wall the railroad tracks ran along the shoreline north of Lakeshore Road. They filled in all the property out between Lakeshore Road and the sea wall with sand from the lake. And that gave a measure of employment."

"The winter of 1909 was a very desperate. The municipality had set up soup kitchens to care for their unemployed. It was surprising what happened. We can look back and laugh now. The soup didn't appear to have very much substance to it. The soup was cold like what was sold as vichyssoise in our finer restaurants, which is considered a delicacy. Every effort was made by some of the local unions to try and hold their membership.. Work was real down, it was real bad. In fact, I had a few weeks of unemployment myself. Ernie Drury, the business agent for the electricians, he started organizing the unemployed. The Labour Temple gave over (rent free) their hall to meetings of the unemployed."

"The hall was full of people, there were union and non-union workers who were out of employment. We used the chance to speak to them to advocate political activity on their part. Later on a little work opened up and the meetings were discontinued. Although we still continued our Sunday night meetings - - they had a good response. Ernie Drury conceived the idea of holding an unemployed parade. Then we couldn't get a permit to hold the parade and we decided to hold it without a permit. It was surprising the turnout. We met down at Bayside Park, that is the bottom of Bay Street, just below the railroad tracks."

"We started to march up Bay Street. Black Jack Robinson, the editor of the Toronto Telegram was standing at his window, he had been one of the main instigators of taking police action against the protesters, the unemployed. Evidently, he tipped the police off or there was some connection with them in my book, cause the police came out of Melinda Street, walked right into our parade.{The Toronto Telegram building was on Melinda Street} They grabbed Drury, Cheeseman and Watkinson and ran them down to Court Street station. They charged them with conducting a parade without a permit."

"Myself, Alex Lyons and Davey Crombie were rushed up to Dundas Street police station. We were given the third degree and at that particular time it was pretty severe. However, they had nothing on us because we hadn't organized it, we were only taking part in the parade. The inspector at Dundas Street let us go with a warning. But later on, they fined Drury, Cheeseman and Watkinson of the carpenters twenty dollars for holding a parade without a permit. But the police had broken up our parade. However, the parade had caused a lot of ill feeling in town and we held a couple of mass meetings in the Labour Temple, organized in protest." (The largest hall in the Labour Temple held between 400 and 450 people. In 1909, that was deemed to qualify as a mass meeting, to some, it still does to-day.)

"Unemployment was still rising and a lot of our people were suffering. Jimmy Simpson and myself were called by Goldwin Smith* to meet him. We went up to meet Goldwin at the Grange." (This old mansion later became the Toronto Art Gallery. Additions have been added to the the Grange, the original Toronto Art Gallery. (Goldwin Smith was a professor, historian and author. His book on "The American Revolution" disturbed Woodrow Wilson, , then a professor, later the President of the Unisted States. Wilson wrote a scathing criticism of Smith's book, a thesis of a person intoxicated with patriotism.)

This is John's report of his meeting with Goldwin Smith, "He said to us, a number of your boys must be suffering very badly. We said, 'Yeah'. He asked, are you prepared to set up an unemployed fund?' If you will, I'll give you a thousand dollars to start it, on the condition you don't tell where it comes from because I am trouble and might incur some displeasure from the authorities for giving you money. However, he donated a thousand dollars."

"The Trades Council called a special meeting to deal with the problem of unemployment. The Trades Council gave five hundred dollars and decided to set up an unemployed fund. We got donations from Mr. Atkinson of the Star and a lot of well-known people. Jimmy Simpson became Chairman of the fund and I became the Secretary. I had to find those who were suffering the most. We then sought funds from the local unions. One must give credit to some of the trades, particularly the bricklayers, they gave us a good donation. And the printers gave us a fair donation. By the end of the year we had pretty near five thousand dollars. The demands on us were so great we tried to appeal to the city hall to give us some assistance. But they couldn't give us any assistance from the government."

Speaking of this time, John said, "The Building Trades Council had proposed building a bridge (at the east end) of Bloor Street, from the end of Bloor Street to Danforth Avenue. Gerrard Street was the northern bridge then. That was the farthest north you could go to get over the Don (River). But we (the Building Trades Council) got no response for that until later. We kept up the agitation through the building trades and the Trades and Labour Council to take on some unemployed workers and give them employment and we thought that was one of the best things that could be done or could happen.

"I was appalled at some of the suffering I found. One particular case I could never forget it, it is indelible in my memory. A little, active member of the labourers' union, named Lawrence, lived on King Street East. He had worked on some of the jobs I had worked on. He lived down near the old car barns on King Street. Came to me at about nine o'clock on Saturday night. He said his wife was very ill and he had six children, no heat in the home. I went down with him and I saw the woman and I was of the opinion she had pneumonia. I got my doctor, Dr. Bryan, who looked after my wife; it was around about ten o'clock Saturday night. He came and saw her. We had to make arrangements to get her into the hospital. We had difficulty in getting an ambulance. It was one o'clock in the morning when we got her into the old Toronto General Hospital on Gerrard Street East. She had double pneumonia. Dr. Bryan said if we hadn't got her in when we did she would have died. Then I went back to his house and found that they had no coal; hardly any food in the house. I went to my home and got a bag of coal. It was after one o'clock Sunday morning, I looked for a store who would give us some provisions to help the kids over Sunday. I found a store and got enough to see them through Sunday."

"I saw cases pretty near as bad. I tried to interest the Social Services Council of Canada to go in with us. Jimmy Simpson was very active at that time, in social service work. And we both appealed to Social Services but we couldn't get much assistance out of them, something I could never understand. So, we continued with our own program even into the next year, carrying into 1909 when there was some activity, some employment and things were getting better.

** Goldwin Smith (1823-1910) born in England was an internationally known historian, advocate, editor and publisher. He came to America when in his early forties. For a time he was a professor of English and constitutional law at Cornell University. He wrote many books. His most controversial was "The United States, An Outline of Political History". This book much upset Professor Woodrow Wil-*

son, later the U.S. President, He said Smith's views were "cycnical and impracticable." What annoyed Wilson was that Smith saw behind all the altruistic, noble and high principles proclaimed by all the participants. Goldwyn Smith saw each side had its leaders, leaders trying to retain power and the others seeking to depose and replace them.

When one learns that wars are usually fought for political and economic reasons, he or she sees the end of the myth that is too often touted as history.

When Smith died in 1910, he left his mansion and grounds, 'The Grange" to the City of Toronto. It became the city's art gallery. Today, the original house still stands and is a part of Toronto's well-known art gallery. It was his donation to help Toronto's needy that served to create the Toronto and District Labour Council 1908-09 relief fund.

Chapter Fourteen - The Delegate

In his lifetime, John Bruce no doubt set a record for the number of conferences and conventions he attended as a delegate. During his years in the labour movement he served as a delegate at conventions of The American Federation of Labour, the Trades and Labour Congress of Canada, the Canadian Labour Congress, the International Labour Organization, the Ontario Federation of Labour, the Canadian Commonwealth Federation. {CCF} the New Democratic Party, {NDP} the United Association of Plumbers and Steam Fitters and many others.

He spoke one or more times at these conventions and conferences. At these gatherings he promoted causes dear to his heart. He was not one to hesitate to say what he believed no matter the hostility he might face. His philosophies were many and not always popular with many of his fellow trade unionists. Despite his strong socialist, pacifist views and propensity to promote lost causes, he was popular and held in high regard by his fellow unionists. His election as delegate to the United Association 1908 convention is another example of his popularity with the members of Local 46. Every officer, including the President, the man usually elected as delegate, were senior to John. He was most active in the local although he had only joined the local in the middle of 1906. He was nominated and accepted and was elected the delegate to the U. A. biennial convention in Indianapolis, Indiana. At this time about a quarter of a million people lived there. It had a two and half mile oval race track; was associated with auto racing and was building up the automobile industry.

In this chapter John talks of two conventions that changed

his life dramatically. In 1969 he was ninety-three years old. On this day he was sitting in a large upholstered armchair in his large living room which was overcrowded with furniture. The heavily-lined drapes were pulled closed to keep out the warm summer sun. On the large glass-topped coffee table, sat a reel-to-reel tape recorder with its microphone placed in front of him. He and his daughter, Amy, had lived in this large two story red brick house for nearly half a century. It was located north of Steele Avenue on Yonge Street in Toronto. With the rent he paid, he could have brought the house several times over. Yet, he never owned a house. When this building was sold and John had to move he went to live with his youngest daughter in a house she had bought. It was a comfortable six room modern post WW Two house, about half as large at the one on Yonge Street. Why he never purchased a home he did not say.

He began this interview telling of the first two conventions of the United Association of Plumbers and Steamfitters he attended. These are John's exact words used to describe these two events; the second changed his lifestyle completely. He found the position for which all his previous experiences prepared him for.

John begins thus, "1908 of course was a banner year for myself. I had been elected to attend the convention of the United Association of Plumbers and Steamfitters in Indianapolis, Indiana. Evidently, some of my popularity had spread. Particularly, I had been to Hamilton a lot, helping the building trades in Hamilton. I was speaking to the trades council in a couple of meetings they had held."

"The first day of the convention registered very much in my mind. An outrageous action on my own part, in my judgment particularly as a socialist. When we went into the hall the walls were decorated an awful lot with American flags. There were no Canadian flags, no Union Jack, at that time was considered the Canadian flag."

"The Governor of the State of Indiana made one of the finest international speeches that one could listen to. He commented on the good spirit that existed between these two particular groups, Canada and the United States. We {the Canadians} had about ten delegates there. {there were three hundred in total} When he was through the President called the convention to order. The very first thing I did, I got up and I protested. I said 'Is this an international?' The President said ' yes'. Well I said, 'Then why has the flag of our country been ignored. We have listened to a stringent speech on the friendship that should exist between us, yet there is no evidence here that there are foreigners here.' Well, the President was a fine fellow and he accepted it and gave a very apt reply. He said, "that will be attended to at this afternoon's session and we will have a flag representing Canada."

"During the discussion in the convention I played a very important part in the discussions. I had won a measure of popularity I had never expected. It almost appeared that I had dominated the convention, particularly in the discussions. The president appealed to me not to be too critical, you know, of the administration - - they were limited in their funds and they tried to tell me of their organizing efforts that had been made by the organization. If I {Bruce} would be a little more tolerant it would be all right."

"I didn't resent it but it seemed to me that somebody wanted to sit on the lid. There was at the time the Secretary-Treasurer, the President and one organizer. We had nineteen hundred members in Canada and they had about twenty thousand in the United States.

"When it came for the time for the nomination of officers —

only one full time organizer had been on the road a couple of years at that time. Somebody nominated for me and I stood for election but I got defeated. The big local unions were able to defeat me at that particular convention. Chicago, New York, Detroit, Philadelphia, Buffalo were fairly big organizations."

John Bruce had learned the power structure of the United Association. More than likely with this knowledge of the power structure and with his temerity, he wanted to test it to see how it work ed.

The 1910 Convention of the United Association was held in the Hibernian Hall in Minneapolis-St. Paul. This western city had a population of approximately three hundred thousand and was a major centre for grain, flour and lumber. John Bruce was elected to serve as a delegate. John tells of this adventure in these words -St. Paul in Minnesota. "At this convention evidently I played a more important part. Prior to the convention I had written President Alpine to tell him not to give any opportunity to raise the question of the flag. And he assured me that would be attended to."

"When we got to St. Paul we were to meet in the Hiberian Hall. I was standing on the sidewalk, when a member from Hamilton, {Ontario} who had been upstairs said, 'John, I don't see a Canadian flag.' The place was decorated with Irish flags and American flags, all over the place. I went up to look and I said, 'When am I going to overcome this? This is funny.' I went downtown. I got the police lieutenant to drive me downtown in his sidecar. I then went looking for the Canadian Consul. I asked him for a Canadian flag. He said, 'Well, I don't have a Canadian flag.' Which seemed surprising to me. Of course at that time some of them used Royal Jack, you know. {probably he meant the Canadian Ensign} He said, "I've got a great big Union Jack about thirty or forty foot long.'

"I said will you lend me that?' He bundled it up, that was quite a bundle you know. I brought it up and was sitting in the back of the hall with the bundled-up flag behind my feet. "

"We had another fine address from the Governor of the state, and he alluded to the wonderful relations that prevailed and all like that. So, after he was through, John Alpine banged the gavel and said "This meeting is open for business.'

"And I said, 'Mr. President, —' . Immediately he turned and looked. He knew what was coming. I said, 'There is no Canadian flag and you promised.'

'Brother Bruce, the letter I wrote you stated there would be one here. I can't understand—.'

"I said, I got one here."

"He said, 'If you bring it forward I'll have it hung,'

"And by god, it covered the whole back of the rotunda, the platform. Not big but massive and cumbersome. Here I go up, the crowd just cheered madly." {Bruce laughs as he tells of this episode.} You know how the flag appeals to them in the United States. "

"The funny part {John begins laughing again} the guy who had the order for decorations was kept waiting outside the door so he could put a Canadian flag up."

Bruce's description of this event was more amusing and more colourful than the verbatim convention report. The official account of this episode fails to convey the humour and commotion this digressive tactic of Bruce created. The slight embarrassment of President Alpine is evident from his response to John's complaint.

This is that verbatim report from the Monday morning session, September 19th, 1910 of the proceedings of the United

Association of Journeymen Plumbers, Gas Fitters, Steam Fitters and Steam Fitters' Helpers of the United States and Canada, convened in St. Paul, Minneapolis. {Page 66}

Delegate Bruce,- Toronto, I would like to call attention to the omission as I called the same attention in the Indianapolis convention. I refer to the absence of the banner that floats over the brothers of the Dominion of Canada from the decorations of this hall. We had distinguished visitors here this morning who paid tribute to the brothers north of us. They spoke of an imaginary boundary line that separates the U.S. from Canada. We feel, however, that we should pay tribute to that flag that floats over us from the north. The delegates from Canada have secured such a banner, and we hope those who have charge of decorating the hall will see that it is put in position.

President Alpine - I am very sure, Delegate Bruce, that a mistake has been made, and that it is unintentional. Some weeks ago, remembering the omission at Indianapolis, I wrote the Committee on Arrangements requesting that there be no repetition of the Indianapolis incident. I was assured that ample provision had been made for such a flag. I am now advised that the emblem of the Canadian Dominion here has not yet been placed in position. I think Brother Bruce is justified in calling our attention to the omission. I had noticed it myself."

One had to hear John speak to appreciate what an effective speaker he was. He was a natural, his resonant voice, his choice of words, his when advisable, impassioned delivery gave his words an evangelistic appeal.

When he cried, "Mr. President, You promised." the listener could readily imagine a four-year old teary-eyed little girl, head down in her disappointment, whispering, " Daddy, you promised." A promise broken, the accused so obviously guilty. Bruce stood silently letting his accusation have its full impact. President John Alpine, defenseless, answered the charge with, "'Brother Bruce, the letter I wrote you stated there would be one here. I can't understand—.'

When John said, "I got one here." it was as if one could hear the blast of a bugle, the rattle of bridles, saddles and stirrups and the sound of galloping cavalry coming to the rescue. The ominous, quiet tension was swept away, harmony and unity was restored. John had saved the day. John with his sense of drama, his impeccable timing and careful orchestration made him 'the man of the hour' If President Alpine had had a Canadian flag hanging in the hall, John would have stood up and thanked him for acknowledging the Canadian presence. Either way, John was a winner. He saw an opportunity to make a greater impact and he took advantage of it. This ability to adapt quickly to a sudden change in a situation served him extremely well all his life.

John continued with his report on the 1910 convention of the U.A.. "However, that seemed to create a condition that I was accepted in our organization in a manner that no one before had ever been accepted. Whenever I took the floor I seemed to be able to dominate the convention. It was getting a little bit embarrassing, you know. I was cheered, you know, if I said anything. "

John was ninety- three years old when the tape was made yet he seemed still a little in awe of the impact he had made on this convention. No doubt, he was the best and easiest speaker to listen to present. He was one of them, a rank and file member and they were looking for a champion. A man ready and able to stand up to the leadership which most of them were in awe of.

Back to John. "There was a resolution before the convention to put on four organizers. The Constitution Committee brought in a recommendation that only one more be appointed, Bill Lynn, that was an additional one. They outlined the expense and all like that. What it would cost, and what position we were in. We were broke then. However, I led the fight and we defeated the committee's report. With the result that I moved that the original resolution become the matter before the convention; that four organizers be put on and that carried overwhelmingly."

John's memory at ninety-three years of age, usually so near perfect failed him in this instance. He was correct in most of the details but he had the numbers wrong. There had been six general organizers, not one, as John said. His speech and resulting amendment increased the number to eight. Nevertheless, his speech swung the balance against the recommendation of the committee and the wishes of the General Executive Board. The following is a summary of the **session and the verbatim speech of John Bruce.**

From the Monday morning, September 26th, 1910 session of the proceedings of the United Association of Journeymen Plumbers, Gas Fitters, Steam Fitters and Steam Fitters' Helpers of the United States and Canada, Convened in St. Paul, Minneapolis. {Pages 124 to 128}

Resolution No. 167

By Local No. 46

Whereas through the inability of the General Office the last few years to give to the Dominion of Canada that attention that we believe is necessary for the organization of our craft and the perfection of the International trade union movement; and

Whereas; that the lack of such organization is leaving that

territory a menace to the U. A. and offering many opportunities to our enemies be it;

Resolved That the Dominion of Canada be constituted a separate district of the U. A. and that an organizer be elected to the said district.

Chairman Clark; "Your committee after carefully considering the above resolutions deem it inadvisable at this time to increase the number of organizers, owing to the fact that our present finances will not maintain it, and recommend that six organizers be elected and assigned to such localities as in the opinion of the General President, the General Secretary-treasurer and the General Executive Board services are necessary."

A motion was made and seconded that the recommendation of the committee be concurred in.

{Several speeches were made in opposition to the committee's recommendations. First to speak against was Delegate Leonard (E. W.) followed by Delegate Higgins, Delegate Malley, Local 120 and several others.}

Among the last to speak was John Bruce whose amendment to increase the number of organizers from six to eight was adopted. A special organizer for the Dominion of Canada was not included in the Bruce amendment.

The following is a verbatim report of John Bruce's speech.

Delegate Bruce — We hear today on every hand that the staff of organizers should be increased. It is reasonable to expect that increasing the number of organizers will bring results. It has in the past. Today there are twelve appeals from different sections of the country asking for assistance. Even with the increased number of organizers, the field is not covered. You will find the members of this

Association have increased to the extent of a little over 2,000. There is reason to think the action of the Indianapolis convention was for the best interest of the United Association. In Indianapolis the funds of the Association amounted to $47,000 as against $74,000 in the St. Paul convention.

Speaking as one for the Northern district, I am going to stand before you and make an appeal for consideration from that territory. In the Indianapolis Convention President Alpine took the floor and told you the I. A. (International Association of Gas, Hot Water and Steam Fitters) was a misnomer that there was no such thing as the I. A. in the Dominion of Canada. I am sorry to say that today in the city I represent we have one of the most disturbing elements that ever occurred there, the formation of an I. A. Local. I am sorry to say also, that my co-delegate at the Indianapolis convention helped form it. Through the inability of the general office to respond to a call, President Chinchilla was able to walk in with a charter in his pocket and induced twenty- two men in the railroad shops to form an I. A. local, after they made an application to the U. A. for a charter. They are using their influence with the disrupted membership in Winnipeg. The delegates from that city say it is the hardest job in the world to fight the injunction proceedings and all that is hanging over them and keep their men.

In Indianapolis I made an appeal to the convention for more consideration for Canada. I am a broad-spirited trade unionist. I do not recognize any line of demarcation. There is no identity between the laws of the two countries but economic and industrial affairs there is an identity of interest among the workers. In 1906 when the convention was held in Toronto there were almost 2,000 members of the U. A. in the Dominion of Canada. In the Indianapolis convention the membership had fallen off, and I am sorry to state that there are only sixteen delegates in this hall representing a little over 1,200 members in Canada. It is evident we are not receiving the attention we should. We cannot boast of the locals that have been added to the U. A. and we cannot boast of the membership we have added to the U.A. Why? If you add up all the time of an organizer in the Dominion since the Indianapolis convention you cannot make two weeks of it. If that territory is of no use to you it is up to you to say so; but if you are prepared to give us some support you must at least vote down the recommendation of the committee and increase the number of organizers. In Calgary an organizer was sent for. They had some trouble. Because the organizer went there he was able to make one shop 100 per cent. That thing rebounded to the credit of the U. A. in Western movement and today all the delegates representing the Canadian districts are pledged to support a notion to place an organizer there.

We realize that if it is impossible for the U. A. to say one man shall be stationed there should be conversant with the laws of the territory. During 1908 a bill was introduced to the Dominion Senate to outlaw International organizations. That bill was defeated. During 1909 a bill was introduced into the Dominion Parliament that we outlaw any international organizer coming in there during times of strife. If that had been adopted those men should be deported. That bill was shelved. Brother Alphonse Verville would say, were he here today, that it was only a subterfuge and that the bill will be revived. The purpose there is to substitute for the International trades union movement a purely Canadian movement and that means nothing but a gang of scabs and strike breakers. They are trying to force on us a narrow Canadian movement, but we are not going to give up the International movement. We wish to stick to the International movement, and we want you to assist us in doing so. During the last two or three years the secretary of my local union was flooded with letters from secretaries all over the country asking about men who have worked in our locals and asking what we know about them.

Most of you know that at the Indianapolis Convention I tried to appeal to the delegates and show that we were not receiving the support we should have. At that time we had 670 members on the books of the Association. Today the largest local is represented by one delegate. Had we been in a position to fight, had we been able to keep a man in the field I believe the I. A. element in our city would not exist.

I move as a substitute for the report that, in the best interests of the United Association the staff of organizers be increased to eight {Seconded by Delegate Ryan}"

Other speakers including President Alpine spoke after Bruce. Alpine realized that those speaking against the committee's recommendation not to increase the number of organizers would carry the day. He saw that Bruce's substitute amendment was the best compromise and so ruled that it was in order.

The substitute offered by Delegate Bruce was carried "The president came down to me and said, 'You bugger, you beat us. You have to be one of the organizers.'

"I said, 'If you think I'm looking for a job you've got the wrong guy. I have already got a new job offered me right now, to go and work here for an American contractor. One of my friends from Toronto is a foreman in Chicago."

"However, they invited me into a caucus and I refused to go because I have always been strongly against machines in politics. I believe in maintaining a measure of freedom, freedom of action and freedom of thought. I refused to go into a caucus. The President appealed to the Canadian delegation to go in the caucus so I could be nominated on their slate. However, I wouldn't go, I didn't go into the caucus but they put me on the slate anyhow. I was on the slate to be elected for Canada. And when the election came there was a Secretary who had been deposed. Because of the actions in the office, you know, the Executive Board had deposed him and had put another person in his place. A fellow named John Lowry of Philadelphia. He {the deposed Secretary} decided to run against me but I got an almost unanimous vote, you know, at the convention."

"President Alpine who wanted to nominate me for the position stepped aside to let a brother from Winnipeg nominate me but he got out of the chair and seconded my nomination. However, as I said I was elected."

"When the elections were through he {Alpine} said, 'Here we go, you're going to go right on the job on the first of October.' You see that was September. 'We can't wait until January we got to get someone into Montreal. There you are, here's a thousand bucks you can take to Montreal- - you're going to be responsible for it. You better go into Montreal and see what you can do to aid them in Montreal."

"With that, that started my career in the United Association of Plumbers and Steamfitters began."

The manner in which John reported President Alpine's comment, 'You bugger, you beat us. You have to be one of the organizers.' it appeared he thought he had been paid a compliment. This may have in part been true. Yet, one cannot help but think the president was telling John, it was for him to put up or shut up. If organizing was so necessary, such a priority, then get out there and do it.

John with the zeal of a missionary had sold the delegates on their obligation to organize the unorganized plumbers. John's physical appearance, his height, stance, wide black moustache, his carriage gave him advantage over others. His elocution training, his timing, his clear and correct enunciation and his ability to read his audiences made him a dynamic speaker. As a boy he listened and learned from the missionaries making appeals for donations— donations to make Christians out of black heathen in Africa. He was impressed with their success in making people give and feel good as they parted with their hard- earned cash. Defeating the leaders of the U.A. may have made him a hero of the rank and file delegates, but did not endear him to the officers. Despite his holier-than- thou attitude to machine politics, the union could use Bruce's several skills. Without John appearing to recognize it, Alpine had made him a part of the administration, so he became part of the political machine. President Alpine knew better than John what had been done.

When asked what was your salary in those days, John answered, "Eighteen hundred dollars a year and three dollars a day, no, fourteen hundred dollars a year, pardon me, and three dollars and a half a day expenses. And we had to account for every little bit of postage and had to make a detailed report of where you went and all like that. Three dollars and a half a day, I can tell you quite frankly, in Canada, we could get by on it. You could live on three and half dollars them days. You could get a place with a lovely room for a buck a day or six dollars a week. Some places five dollars a week— in some good hotels. Meals, you could always get good meals for thirty or forty cents. A good meal for forty cents. Now you pay that for a slice of bread."

"I left the convention and came home and got my affairs

straightened out. My wife said. 'You're out every day and night and Sundays included and you might as well get paid for it."

One must wonder, had the delegates known that they were electing a pacifist, a Marxist socialist, a strong temperance advocate and an agnostic, would they have done so. John's eloquence, his skill as a labour evangelist and his cautious promotion of philosophies served him well. He had the talents required to be a union messiah, a missionary spreading the gospel of unionism. He was a bitter opponent of religious dogma, bigotry and found much fault with the clergy. Union constitutions, for good reason presribed, nonsectarianism and John abided by this rule. "We had very few of our cities organized at that time. Halifax, Montreal, Ottawa, Kingston, Hamilton, Toronto, Winnipeg, Regina, Vancouver and Calgary. They were the major centres at that particular time."

Chapter Fifteen - The General Organizer

On Friday, September 16th, 1910, John Bruce put down his tools, never again to thread a pipe, tighten a Stilson wrench on an elbow, tee or couplings or sit at noon with his fellows on the job site with a lunch pail on his lap enjoying the camaraderie of a man working at his trade. His days as a journeyman plumber were over. He had a few days in which to rearrange his life; give his employer notice; turn over his various union duties to other members of Local 46. Although he was leaving behind his Toronto Local to represent all the men in his trade in Canada. All his life, he always saw himself as a member of Local union 46 of the U.A.

On October 1st, 1910 he began his career as General Organizer for the United association of Journeymen Plumbers, Gas Fitters and Steam Fitters' Helpers of the United States and Canada. His first assignment was to attempt to settle a strike of Plumbers in Montreal, Quebec and Ottawa, Ontario.

When John Bruce began his career as a General Organizer he was thirty-five years old. He was five foot ten inches tall, three to four inches taller than most of his fellows. He was of a sturdy build and sported a lush black moustache. He always took pride in his appearance and wore well-tailored three-piece suits and a black derby hat. His dress, posture and self-assurance set him apart from those around him.

For most of his years as a General Organizer John traveled by rail. From Toronto to Winnipeg it took two days and four days to reach Vancouver, B.C. On most of the trips he broke his journey by stopping over at towns and cities along the way. He traveled from city to city. town to town by train, he would cross the country and do so an average of four to five times each year.

In addition to meetings with officers, shop stewards and business agents and attending membership meetings he made side trips. Side trips to visit towns and cities to check out organizing possibilities. He was a tireless organizer. No matter what other work he was engaged in he never passed up an opportunity to recruit new members; organize a new local union. In his first four years he doubled the membership of the United Association in Canada.

Until commercial airline travel was readily available he traveled from Victoria, British Columbia to Halifax, Nova Scotia by train. It was the practice for general organizers of the various building trades to journey together in groups of three or four. John might travel with the General Organizer of the Carpenters and Joiners, the Electricians, Painters, Boiler Makers or the Iron Workers' Union.

For more than half a century John spent more nights sleeping in hotel rooms than he did at home. He spent very little time with his wife and family. In 1910 John's salary was fourteen hundred dollars a year and his per diem expense allowance was three and a half dollars a day. He kept a list of hotels which he frequented and the price they charged in his diary. In his 1913 diary all the hotels he patronized, save one, charged a dollar a day. The daily rate of Walper House in Kitchner, Ontario was one dollar and fifty cents. If John booked in for a week or more he got the reduced weekly rate of five or six dollars. John said the per diem he was allowed was quite adequate as a good full course meal could be bought for thirty to fifty cents.

Except for a few months, John filed a report of his activities every month which was published in the 'Plumbers, Gas and Steam Fitters' Journal. There are nearly six hundred of them. In those reports he was careful to avoid putting a member or local union in a bad light. When he was compelled to audit local union account books which was required of him from time to time, he rarely reported what was wrong nor name or criticize the person responsible for any shortage or misuse of funds. As the dues were collected in cash and as many members paid in coin twice a month, it was a difficult task to keep an accurate record of who paid and how much. Members delinquent in the payment of their dues would on occasion, be exonerated from paying the full amount of the dues owing. This only added to the problems of the Financial Secretary and Treasurer.

John, traveled extensively, speaking to thousands of people. Those who heard him were impressed by his splendid oratory and convincing logic. By his own admission he was a dynamic speaker. Some said that he was a laid-back evangelist. He had a great stage presence. This marked him as a true crusader and advocate of union and people's rights. It was skill that he had acquired at an early age. He welcomed the opportunity to expound his socialist and humanitarian views.

He was at times sent out of his own region by the International President to deal with some difficult, potentially explosive political problem in local unions in Buffalo, Detroit, Seattle and other cities in the United States.

He tended to go 'overboard' in condemning unions who infringed, or whom he thought infringed, on what he deemed to be the special domain of the Plumbers' Union. He was especially vehement in condemning rival plumbers' unions. John was never hesitant in condemning their leaders as being a bunch of scoundrels and knaves. Their members had been duped by the false promises of their leaders who were masquerading as trade unionists.

In 1910, Local 78 of the International Association of Gas and Steam Fitters represented the majority of the men in this trade in Toronto. John branded this local as an 'outlaw outfit'.

In 1912, this 'outlaw outfit', the International Association of Gas and Steam Fitters merged with John's own union to become the mighty International Association of Gas and Steam Fitters United Association of Journeymen and Apprentices of the Plumbing & Pipefitting Industry of United States and Canada "

In October, 1910 when John Bruce began his long career as a General Organizer, the United Association had 16.664 dues paying members in 525 local unions in North America. There were in Canada. 1,571 members in thirty- two local unions spread throughout the provinces in Canada. .

(Estimated)

Membership	**Province**	**Number of locals**
125	Nova Scotia	2
100	New Brunswick	1
220	Quebec	2
650	Ontario	14
95	Manitoba	3
100	Saskatchewan	3
90	Alberta	4
210	British Columbia	3

In 1910, Canada's population was 7,207,643. Many people

were moving to western Canada where land was cheap and there were greater opportunities.

The U.A. mechanics working for the railroads were an important part of this group. They were instrumental in moving the membership in western Canada.

Source The Canadian Government publication, Labour Organizations in Canada. 1911

It listed 32 local unions of the United Association in Canada. According to John Bruce's monthly reports by the end of 1911 there were thirty-four.

By the end of November, 1911 John Bruce had met at least once with the officers and members of each of the 34 local unions of Plumbers in Canada. The "State of the Local Unions in Canada", has been compiled from his monthly reports published in 'Plumbers', Gas and Steam Fitters' Journal".

He was elected in September, 1910, and began working as a General Organizer on October 1st, 1910. His official term did not begin until January 1st, 1911. At this time he was a 'new boy' and admitted he was '"green grass". During the months of October, November and December 1910, he was involved a strike of plumbers in Montreal and Ottawa.

In 1911, four new local unions were organized. They were Sydney, N. S., Moncton, N.B., Berlin, Ontario and Windsor, Ontario. The local union in London, Ontario was dissolved.

Source -Labour Organizations in Canada, 1911 edition

STATE OF LOCAL UNIONS IN CANADA

Local Union	City	Population -1910
46	Toronto	376,538

This local was still recovering from the year long strike of

1907. To add to its problems Local 78 of the International Association of Hot Water and Steam Fitters represented the majority of the men in this trade. Bruce said that Local 46 was making "strenuous efforts to rebuild its strengths." Bruce and Business Agent Storey were secretive about their efforts to steal members of Local 78 of the I.A. for Local 46. Bruce and Storey were in constant contact with the members of Local 78 of the I. A. Bruce branded this local as an "outlaw outfit". During the year former members who had joined Local 78 of the I.A. began returning to Local 46,

Canadian Pacific Railway was building a fifteen-story office building at Yonge and King Streets. It's estimated price of construction was $900.000.00. Toronto opened bids for a new sewer system. For most of the year building trades men were employed.

In 1911, Bruce found the indifference of many old members difficult to understand. There was a large number of men working under any condition for the sake of having a job. Men were receiving the same rate as they had four years ago. Bruce blamed the International Association of Steam Fitters union and said the reason was there were so many unqualified men engaged in the trade.

The average hourly rate for Toronto plumbers {Union-and non-union} in 1911 was forty cents.

56 Halifax 46,619

Conditions here were not good. Although the building code was a good one, it was not enforced. Bruce found that the steam fitters were receiving the lowest wage rates in the country yet they were not prepared to unite with Local 56. Bruce believed that soon the members would take a keener interest in the affairs of the local.

Plumbers' and gas fitters' work was slow. The average hourly rate for Halifax plumbers {Union-and non-union} in 1911 was thirty cents.

67 Hamilton 81,969

Bruce found this local was going strong and the members were out for better conditions. He believed that they would be successful in this endeavour. At its annual banquet on March 24th, 1911, Alderman Clark, an old member of Local 67, represented the master plumbers, spoke and was well received. John Bruce was also a speaker. All the building trades were busy.

71 Ottawa 87,074

This local was having problems with Garth Co., a Montreal firm, that was building the Chateau Laurier Hotel. The contractor was enforcing Montreal conditions upon the mechanics, which were inferior to those rates paid in Ottawa. When Local 71 presented the Ottawa agreement to Mr. Meadowcraft, General Manager of the Garth Company, he refused to accept it. Local 71 placed the Garth Company on the unfair list.

Late in that year the local was recruiting new members and five new members were initiated at the November, 1911, membership meeting. Still Bruce was displeased with the lack of interest in the local, despite the fact that the members enjoyed the best conditions as existed anywhere in Canada.

There was a shortage of carpenters but there was work for all the other building tradesmen.

144 Montreal 470,480

John Bruce sent into Montreal by President Alpine, had held meetings with the Chairman of the Plumbing Contractors to no avail. The local had ended an unsuccessful strike in December, 1910. Bruce's view was that "The conditions at this time of the year were far from being ideal and the preservation of our organization was the main factor," The plumbers had ended their strike although no settlement was reached, Bruce reported that the work stoppage had forced the employer to increase wages 5 to 15 cents an hour. Bruce did not confirm this statement when talking of the Montreal Strike in 1969.

The Carpenters and Joiners struck for better wages and conditions. The large employers agreed to pay more wages but refused to recognize the union.

Because of the strike more men were working than was usual in the winter months. In October, 1911, Bruce found the members apathetic. In his speech at the October 17th banquet he urged the members to take a greater interest in the union and become more active in promoting the U. A.

The average hourly rate for Montreal plumbers {Union-and non-union} in 1911 was twenty-eight cents.

170 Vancouver 123,902

Two large shops in this city had been listed as being "unfair". Fines were levied against members of Local 170 who still worked for these two contractors. This made matters more difficult. The general strike in 1911 of the building trades had failed to make any improvements. It was a turbulent year for the building trades in Vancouver.

The average hourly rate for Vancouver plumbers {Union-and non-union} in 1911 was sixty-three cents.

179 Regina 30,213

The local union was still recovering from their 1910 strike. The members hoped to get an acceptable settlement and strike if it were necessary. In order to do so, President Lewis and Secretary Smith, officers of the local, were campaigning to restore its strength. The town was suffering from the failure of the slack enforcement of the building by-laws. The local was demanding more rigid enforcement of the building code. The result was many sub-standard buildings were erected that would soon need major repair.

The work for the bricklayers and masons was slow but the plumbers and carpenters were busy.

186 Brantford 23,132

With Secretary Croucher and President Smith, John Bruce approached the city council to appoint a practical man as a plumbing inspector and warned of the unsafe sanitary conditions in Brantford. This campaign continued during the remainder of the year. John was pleased with the officers of Local 186. All the building tradesmen were well employed.

204 Saskatoon 12,004

In this year, all the members were working. Bruce was pleased with the leadership provided by President Thompson and Secretary Brown.

221 Kingston 18,874

This local was not active enough to suit John Bruce. He believed that in time this local would be "up and going". When there in November 1911, Bruce was not pleased with the little progress the local was making. The lack of an adequate building code was hurting the trade. The local was not working hard enough to obtain better legislation.

The contractors and others complained that help was scarce because the workers were emigrating to Western Canada. All the building tradesmen were well employed.

244 St. Catharines 12,484

Bruce found the local in good shape and conditions in the district to his liking.

254 Winnipeg 136,035

This was the local to which the plumbers in the building trade belonged. {Plumbers working for the railroads were in a different local union.} There were many non-union men working in this city. Bruce, with the help of Brothers Cooper and Boyd Miller, talked to those who were working and were not union members. This local was not making as much progress as great as some thought but Bruce found its progress acceptable.

With the Secretary of the Local, Bruce distributed literature and contacted many mechanics outside the U.A. On August 25th, 140 men attended an open meeting. Bruce reported "a good many men made application for membership."

Despite the successes, Bruce reported he found the conditions around Winnipeg deplorable and many of the members apathetic. The influx of non-union mechanics who were ready to accept sub-standard wage rates was a major problem. Bruce hoped that the good leadership in the Winnipeg locals would put the matter right.

The average hourly rate for Winnipeg plumbers {Union-and non-union} in 1911 was fifty cents an hour.

257 Fort William Not Available

This was an active local and Local 378 had succeeded in obtaining an agreement covering this town and Port Arthur. Local 379 and this local were working under a mutual agreement and "things are very good here." The local held its annual banquet on October 3rd. All the building tradesmen were well employed.

264 Saskatoon 12,004

Things were very good in this city. All the members were working and prospects for the future were excellent. Bruce was well satisfied with the work of President Thompson and Secretary Brown.

292 Montreal 470,480

This local was waging an energetic campaign to recruit more members in the railroad shops. They were intent on building enough support to get a better deal from the railroads. Bruce believed that all the workers in the shops would, in time, join the U.A.

324 **Victoria** **31,660**

This local was enjoying very good conditions. Almost every member was employed. Bruce was somewhat concerned with conditions that prevailed on the west coast and urged the members to work to improve their working conditions.

332 **Winnipeg** **136,035**

This was a local of steam fitters. This local was in negotiations with the contractors at this time. There were many non-union men working in this city. Bruce, with the help of Brothers Cooper and Boyd Miller, talked to those who were working who were not union members. Bruce took steps to headoff any attempt by the International Association of Steam Fitters to raid the membership of this local. The average wage rate for plumbers, union and non-union was fifty cents in 1911.

347 **Quebec City** **78,190**

Bruce found this local in poor shape and changes would have to be made. The conditions were very bad and the members were poorly paid. He had a problem as he spoke no French and most of the members spoke no English. John's evangelical speeches had little impact in Local 347. Secretary Jackson was doing his best against considerable odds. Bruce thought in time this local would give a good account of itself. Plumbers were in demand and all were busy.

348 **Lethbridge** **8,050**

With Secretary Wilson, Bruce visited various sites gathering information on conditions in the trade. Things were pretty quiet in what Bruce describes as a promising town. The members, having survived the strike, were very hopeful of the future.

361 **Peterborough** **18,360**

Bruce said that this local 361 would be one of the most progressive locals.

378 **Port Arthur** **11,220**

This was an active local and along with 267 had succeeded in obtaining an agreement covering this town and Fort William. On October 3rd, all the members attended a meeting which was evidence of the good work of President Bell, Secretary Dennis and Executive Member McRoberts. At this gathering, Bruce reported on the status of the U. A. and gave one of his inspirational "go get' speeches.

Plumbers were usually able to find employment.

397 **Sydney** **17,723**

This was one of the new locals. All the men in the domestic line were doing well. Two new members were initiated at December 14th meeting.

This town was the home of the Dominion Steel Company which supplied many unfair employers in the plumbing and steamfitting trade with a good many men. The steel company had a system of espionage and a large police and detective force that would make it suicidal to attempt to organize the pipe fitters. Bruce felt that only if all the trades worked in harmony could Dosco be unionized.

414 **Sault Ste Marie** **10,984**

This local was intent on obtaining better conditions and there were strong indications that they would be successful. It was a growing town and Bruce believed it would become Canada's Pittsburgh, There was a labour shortage in this city.

468 Calgary Not Available

This local was composed of railroad pipe fitters. This was one of the small live locals. Bruce thought they were a militant group. They worked for the C.P.R.

479 Winnipeg 136,035

Members of this local worked for the Canadian Pacific Federation. The local had "put up a gallant fight won out over the Canadian National Railway. Local 479 began action to obtain the same new benefits from C.P.R. They were successful in establishing a new wage scale for C. P. R. western lines that gave them a four and one half cent wage increase. Unfortunately, a good number of the railroad steam fitters belonged to the Railway Carmen's Union. Steps were undertaken to persuade them to join the U. A. John never admitted he raided another union but was merely moving the members to the union to where they belonged, the U.A.

The average wage rate for plumbers, union and non-union was fifty cents in 1911.

488 Edmonton 24,704

The building contractors were determine that they operate "a non-union/open shop". Local 488 and the other building trades were uniting to fight for a 'closed shop'. Plumbers had almost full employment.

495 New Westminister 55,679

Conditions here were fairly good and all the members were working.

496 Calgary Not Available

The local had signed a two-year agreement with the Master Plumbers. The city council, at the urging of Local 496 and the Master Plumbers, introduced a by-law that made it necessary to use licensed plumbers to install plumbing, sewers and drains. It was working out very satisfactorily. This local was showing wonderful advancement. Its membership rolls had risen from 125 to 248. Calgary was growing very quickly which helped explain the greater than usual membership increase. Bruce said, "Calgary was going to be one of the great towns in the Northwest Territory." Bruce gave Business Agent Laraway credit for much of this local's success.

527 Berlin (Kitchner) 15,484

The new members were determined to make the union bigger and better. Every member attended the March 8th, 1911, meeting when the new officers were installed. As so often was the case, John did not say how many attended. He reported that there was a shortage of carpenters.

531 St. John, N. B. 42,511

He discussed the situation with Brothers Hennebery, Burns and Quinn, and concluded "the town was not in good shape." Low quality work was being done as the inspector was incompetent. There was cheap labour and boy labour. Yet all but two or three of the mechanics were union members Bruce found the members had the right spirit and he expected things to get better very soon. The local was pressing the city council to enact a building code that would protect the health of the people. Bruce believed that their efforts in this campaign would soon bear fruit. Conditions were fairly good and all the members were working.

548 Moose Jaw 13,823

In August ,1911, a local was formed when 18 men signed an application for a charter. Temporary officers were elected. On September 4th the officers were installed. Bruce thought the new officers were a very promising group. The local participated in the

Labour Day parade which was spoiled by rain. Still, in view of the bad weather, there was a very good turn-out. The city had very fine building by-laws that were being strictly enforced.

554 Guelph 15,175

Bruce, with the help of Brother Parker, a volunteer A. F. of L. organizer, secured enough members in March, 1991 to apply for a charter and set up a crew of volunteer organizers to recruit all in the trade. A charter was presented on March 30th and the new officers were installed. Conditions were fairly good and all the members were working.

558 Moncton 11,345

The men in the plumbing trade were not being given fair treatment. With the assistance of Brother Bleakley of the A. G. L. a meeting was convened. It was well attended and those present without hesitation joined the U.A. and a charter was applied for.

==

In 1911, the towns and cities were home to fewer people and union membership, as was Canada's population, was considerably less. In Canada more people were promoting better plumbing, drainage and sewer systems. Plagues of cholera and typhoid fever that had killed thousands of people throughout the world and were only recent happenings.

The U. A. by promoting better building by-laws were protecting the health of the inhabitants of cities and towns. It also served to make it advisable, if not necessary, to make sure that those installing plumbing, drains and sewers were properly trained people. It helped to protect the trade as it made certain these workmen were qualified and licensed.

Not all municipalities were convinced that indoor plumbing

with proper drains and sewers were necessary. Some argued that outdoor privies served the purpose and were adequate. John Bruce. a long ardent advocate of good sanitation systems, was constantly encouraging the local unions to campaign for building by-laws that did protect the residents of the town and cities in Canada. John lead delegations that appeared before municipal councils to promote better sanitation.

The following are three reports of his first ninety days as a General Organizer and the strike he was sent to Montreal to help settle. The first are progress reports as filed with the head office and which were published in the November and December 1910 and January, 1911 Plumbers', Gas and Steam Fitters' Journal. The second as he remembered these days in 1969 — almost sixty years later.

PLUMBERS GAS AND STEAM FITTERS' JOURNAL.

REPORT of GENERAL ORGANIZER JOHN BRUCE.

Montreal, Oct. 13, 1910

To the Officers and Members of the United Association:

Brothers, I submit for your consideration my first report and trust it meets with your approval. On leaving St. Paul I proceeded to Toronto to clean up matters that concerned my Local 46. While in Toronto I appeared before the Royal Commission on Technical Education and gave evidence on behalf of the building trades of that city. On Thursday, Oct. 5, I proceeded to Montreal to look after the interests of Local 144, which had gone on strike three weeks previous. The conditions that prevail in this city became unbearable and Local 144 presented its demands to the Master Plumbers' Association. Believing they were not receiving due consideration, the local declared a strike on Saturday,

Sept. 17, and up till the present time have made very satisfactory headway. The Master Plumbers absolutely refuse to treat with the local committee and I am using every endeavour to bring about an honourable settlement. Very many independent employers have signed our demands and we have placed a good many men to work. Most of the large work in the city is at a standstill and we have completely stopped the work on a large hotel at Ottawa, where a Montreal firm has the contract. The situation today is very encouraging and the prospect seems bright for a satisfactory settlement, and I am hoping that the conditions will soon be such that the city of Montreal will be one of the best in our Dominion. The boys are standing firm and I am glad to say we have had very few desert us in this struggle. Trusting the above meets with your approval.

Yours fraternally,
JOHN W. BRUCE
General
Organizer

==================================

John Bruce was eighty-five years old when he wrote his last monthly report to Plumbers' Gas and Steam Fitters' Journal. For a man his age, it is difficult to believe he could be so active, travel so much and be so observant. At the beginning of December 1961 he was in Kansas City, moving on to Lafayette to Chicago then to Milwaukee and home to Toronto. This is last report from John Bruce to appear in the monthly Journal of he Plumbers' Union.

PLUMBERS GAS AND STEAM FITTERS' JOURNAL.

REPORT OF GENERAL ORGANIZER JOHN BRUCE.

December 21st, 1961

To the Officers and Members of the United Association:

In closing my last report, I was attending the twenty-eighth convention of the U.A., at Kansas City. I can speak with enthusiasm of the value of the convention as I look at the years since I attended my first convention in Indianapolis in 1908, which 297 delegates were present. I have attended every convention since that time. It was wonderful to see more than 2,600 delegates present with their wives and friends. This is a tribute to the growing strength of our organization, and the interest displayed by our members, by their attendance.

I wish to express my appreciation to the delegates and executive officers, for again electing me a general organizer. I have served 51 years in this capacity, and hope I may be spared to continue the work that I have done since I first became a general officer.

The general acceptance of the principles and policies, the report of General President Schoemann, which was generously received, and the approval of the work of the administrative officers, all show a great measure of confidence in the policies which have contributed to the success of our United Association.

The address by Secretary-Treasurer McDevitt, national director of the Political Education Committee of the AFL-CIO was an outstanding contribution, demonstrating to the workers the necessity of taking political action, a position I have taken all my life. I submitted a resolution to this effect, to the Indianapolis convention in 1908, advocating affiliation, or the building of a labor political movement. The spirit in which this address was received, and the action, of the convention, in urging our locals to set up political action committees, are a step forward in creating interest in political action by the workers.

Leaving Kansas City, I proceeded to Lafayette to attend the apprenticeship contest at Purdue University. Here was another demonstration in support of the policies of General President Schoemann, in advancing the interests of apprentice and journeyman training programs. These contests create interest among our members in our forms of training and our efforts to give our members opportunities that do not exist in other organizations. I do not intend to deal with the details, which will be outlined in the Journal under reports of the contest. I also attended a number of meetings of the National Plumbing committee, dealing with our program and policies for the future.

Leaving Lafayette, I proceeded to Chicago, where I had an opportunity to confer with Brother Quinn, Secretary of Local 597, and his officers, and old files of the local for records those responsible for the formation of the international association of steam and hot water fitters. I wish to express my gratitude for their assistance.

I then proceeded to Milwaukee and met Brother Enright of Local 601 and his officers, and appreciated the opportunity to search some very old files of records of Local 18. While in this city, which produced the winning steamfitter in the apprenticeship contest, it was a pleasure to meet Mr. George Schilter, the apprenticeship instructor. I had the opportunity to look over the splendid training program and facilities of Local 601, and they are to be commended for their efforts in this field.

Returning to Toronto, I contacted Secretary Coslett, of the Hydro Construction Council re the situation prevailing in negotiations regarding contracts for Ontario Hydro. Later in the week, I attended a meeting of officials of Ontario Hydro, and we discussed conditions to be applied at Grass Island, in the Niagara area. We anticipate an early agreement.

I also attended a meeting of the executive committee and business agents of the Toronto Building Trades Council, in connection with the campaign to organize the residential and apartment house field in this area. The situation is progressing favorably.

I also consulted with Business Representatives Newmarch, Whitehead, Joe Dwan and Secretary-Treasurer Watson on general situation prevailing, and on conditions of the agreement affecting the residential and apartment house field.

I have also been working to set up a negotiating committee for discussion of our new agreement with the Canadian Automatic Sprinkler Association, in behalf of Local 379, and anticipate an early meeting in this regard.

Having received several complaints in connection with some of the sprinkler work in this jurisdiction, I have contacted the companies and have visited several projects. I find that matters are progressing satisfactorily.

I have been spending several days assembling information and conferring with the editor on the work of building up a history of the United Association.

I have also attended a special meeting of the Allied Construction Council dealing with our Hydro problems. These have not been successfully adjusted, and it is the intention of group of international officers to visit the projects and confer with our members who are working on them. I am still working in and around Toronto, and my next report will commence from here.

May I take this opportunity to wish a Merry Christmas and a Happy New Year to all our membership

In 1919 John Bruce was forty three years old. Woodrow Wilson was President of the United States and the League of Nations was a reality. Eamon De Valera was chosen as head of Ireland's new government. People were reading Somerset Maugham's book, "The Moon and Six Pence" and Blasco's Ibanez's "The Four Horsemen of the Apocalypse

John Bruce in the last photograph in which he is seen wearing a moustache. He was in his late forties and seems quite pleased with himself and the world.

John Bruce, the outdoorsman. As a boy he enjoyed hunting and fishing. No sign of fishing gear or hunting equipment can be seen in thi photo.

Jack Bruce as he called himself in 1937 when he ran in the Toronto riding of Davenport as the CCF {Cooperative Commonwealth Federation} candidate for the a seat in the House of Commons. His campaign manager was George Parks, better known in later life as Eamon Park. A *copy of the leaflet cover.*

DAVENPORT RIDING

Vote Labor - - C.C.F.

JACK BRUCE

The voice of the people desires a change.

Make it a real change by voting to change the whole system.

The old political parties inform you that it has broken down and needs reforming. This is impossible under the present system.

Don't trust to promises of reformation. Vote to reorganise it upon a newer and fairer basis.

Why not give the great mass of the people a right to enjoy the comforts of life and not the few? Do you want work or charity?

Listen to the Voice of the C.C.F. This is an appeal to all people to vote for themselves, and not for monopolies and bankers.

The C.C.F. offers relief to a struggling world:

By a system of planned economy.

By a programme of reconstruction and the removal of slums.

By the nationalisation of the banking system.

By finding work for our Youth.

It offers creative opportunity to all. So why not do your share by voting C.C.F.

Vote for JACK BRUCE.

Committee Rooms

276 Havelock Street - MElrose 6500

1201 St. Clair Avenue West

Issued by George Park, Official Agent, 130 Salem Avenue

The Stafford Printers Limited 223 Richmond St. W. Toronto

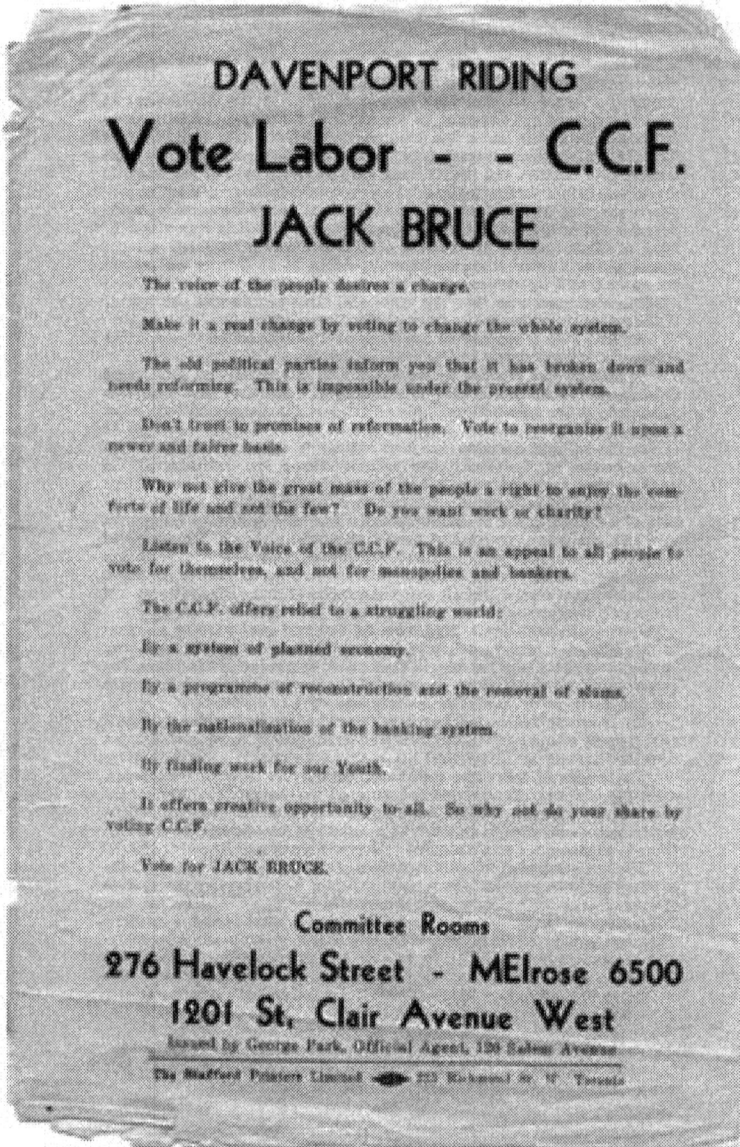

The leaflet used in John Bruce's attempt to be elected as the Member of Parliament for the Toronto Riding of Davenport. His campaign manager was George Park, a University of Toronto student, better known in later years as Eamon Park

John enjoying the warmth of a summer's day. In a shirt and wearing a bow tie. The only photograph in which John is seen wearing a bow tie. At age 57 he may have been attemptiing to change his image.

John still the outdoorsman but a little older on the same steps. No patch pockets on his shirt. The breeches and boots look the same {See page 176}

Plumbers' John W. Bruce Retires After 52 Years as Organizer

John W. Bruce, O.B.E., a general organizer in Canada for the United Association of Journeymen and Apprentices of the Plumbing and Pipe Fitting Industry of the United States and Canada (AFI-CIO-CIC)

John's personal crusade was to have a plaque erected to commemorate George Loveless, the leader of the Tolpuddle Martyrs. After serving his seven year prison sentence in Australia, Loveless settled near London, Ontario. The Honourable Leslie Frost, Premier of the Province of Ontario, asked John to unveil the plaque on April 28th, 1959.

The last photograph of John Bruce taken in a professional photographer's studio. John was eighty-one years old.

John made his last appearance as a public speaker on Labour Day, 1968 when he accepted the William Jenoves Award for his long service to the Plumbers' Union and the labour movement.

John Bruce on his eighty-ninth birthday retired but still busy writing to his friends and colleagues.

John Bruce as he poses for his ninety third birthday photograph. From the February, 1969 issue of "Trade Winds" the publication of Local #46 of the Plumbers' Union {Toronto

Addendum

On September 30th, 1910, when John Bruce put down his tools never again to work at the trade he was thirty-four years of age. On the first of October 1910, he began his long and colourful career as a General Organizer for the United Association of Journeymen and Apprentices of the Plumbing and PipeFitting Industry of the United States and Canada. {Founded on October 11th, 1889 in Washington, DC}

He retired on January 1st, 1963 and died on April 3rd, 1970. This account only covers the first thirty-four years of his life. Yet to be written is the chronicle of his fifty-two year career as an official of the United Association and the seven years of his life he enjoyed as a pensioner.

To help establish John Bruce's place history, it must be remembered that he became a journeyman plumber in 1893, before Australia became a country. Queen Victoria did not proclaim the Commonwealth of Australia until 1901.

When John arrived in Canada, Sir Wilfred Laurier was Prime Minister. Teddy Roosevelt was the American President and when John died, Richard Nixon was U.S. President and Pierre Trudeau was Canada's Prime Minister.

In John's sixty years in Canada there were eight Prime Ministers. After Laurier came Robert Borden, Arthur Meighen, R. B. Bennett, MacKenzie King, Lester Pearson, John Diefbaker, Louis St. Laurent and Pierre Elliott Trudeau.

During John's lifetime, England had 6 monarchs. Queen Victoria, Edward VII, George V, Edward VIII, George VI and Elizabeth II.

During John's years in Canada there were twelve U.S. presidents, Teddy Roosevelt, William H. Taft, Woodrow Wilson, Warren

Harding, Calvin Coolidge, Herbert Hoover, Franklin D. Roosevelt, Harry Truman Dwight D. Eisenhower, John F. Kennedy, Lyndon Johnson and Richard Nixon.

Bruce's career as a General Organizer, covered 52 years of the 20th Century. It is a unique record never equaled by any other union leader. He was a left-wing political activist, a devoted pacifist, a temperance advocate and a Member of the Masonic Order. When provoked he was ready to resort to fisticuffs.

Train travel gave John the opportunity to read more than most. He read every book that Upton Sinclair and H. G. Wells wrote. He devoured biographies of politicians alive and dead. He had been exposed to the gospel of Marx in his early teens and could quote Karl Marx at great length. From time to time he re-read portions of his writings. Yet he opposed the Communist Party which he saw as a great danger to the socialist and labour movements.

In more than half a century as a trade unionist John Bruce rubbed shoulders with the labour leaders and political figures of his day. He was on a first name basis with all three presidents of the AFL/CIO. Sam Gompers its first president.{January 27th, 1850-December 13th, 1924} one of the founders and first President of the American Federation of Labour. John said that Gompers would sit and talk to you for hours as long as you bought the beer.

John didn't see William Green, its second chief officer. {March 3rd, 1973 - January 10th, 1952} as a strong or effective leader. John shed no tears when George Meany, a fellow plumber, replaced Green to become the third of AFL/CIO President. Meany served in that post from 1955 to 1979.

John said he met young George in 1922 at the American Federation of Labour Convention. He described him as brash young Irish

Mick. John was later to admit that George Meany was a most influential union leader.

As the spokesman for the U.A. in Canada John was active in the Trades and Labour Congress of Canada. He worked well with Tom Moore, President of the TLCC from 1918 to 1936, as he did with P. R. 'Paddy' Draper, its long time Secretary-Treasurer. He was no fan of Percy Bengough but could work with him when necessary. Pat Conroy of the United MineWorkers and later Secretary-Treasurer of the Canadian Congress of Labor John considered a well-informed able union leader. Arthur Hemming of the International Brotherhood of Electrical Workers and the last Secretary of the TLCC he thought was adequate but only when second in command.

John and Tom Moore served together on several government boards and commissions and they were members of the first Canadian delegation to the International Labour Organization meetings in 1922.

Of all the unionists John worked with, none was as close or as long a friend and associate as Jimmy Simpson. He was a newspaper reporter in Toronto and worked with John in the Toronto Trade Council and in the Trades and Labour Congress of Canada. Both were avid socialists and devoted trade unionists. Each respected the other and they enjoyed working together.

The leaders of the labour movement in Canada were small in number compared to the United States. They met often and at times quarreled over jurisdiction, political affiliations and union priorities. They had their factions, cliques, and coalitions. These alliances shifted re-grouped fell apart for any number of reasons. Whether allies or opponents, the labour leaders of the day came to know each other very well; maybe too well.

John saw dissidents as enemies whose organizations had to be restructured or eliminated. Their members should be persuaded to become members of unions such as his own. The Canadian Communists at the top of his 'Hit" list were Joe B. Salsberg of the Workers Unity League, Pat Sullivan of the International Seamen's Union of North America and Sam Lapedes of the Garment Workers, plus the lesser known "Commies" .

He described R. B. Russell of OBU {One Big Union} as a misguided well-meaning individual who had lost his way. Still, John asked him to intervene in an attempt to settle the Winnipeg General Strike of 1919. Russell refused to try. He said if he did try he would be thrown out of the fourth story window of the room where the Committee of One Hundred held their meetings. John agreed that "They would you know."

John met James Keir Hardie {1856 - 1915} when he visited Toronto and was much impressed with the Scottish labour leader. Hardie was an officer of the Lanarkshire Miners Union, a socialist and for a time was a Member of Parliament for Merthyr Tydvil. John rarely missed the opportunity to speak well of Keir Hardie and of his few meetings with him.

On a few occasions John met with Ernie Bevan, {March 9th, 1981- April 14th, 1951} a union leader and member for many years who served as a wartime cabinet minister. John found him an interesting and dedicated trade unionist and socialist.

When William Lyon MacKenize King was a young lawyer in Toronto and before he entered politics he and John Bruce spent time together. They shared meals and attended meetings. Most were held in the Alhambra Hall that was located on Spadina Avenue. It may have been named after the Alhambra Hall in London, England, a famous music hall at which the very popular comedienne Marie Lloyd performed for over three decades. One memorable night they went together to hear Emma Goldman the famous American Communist speak. Their association ended when King went to the United States to work for John D. Rockefeller Sr.

In 1908, John and a group of Toronto fellow unionists formed the Social Democratic Labour Party, {which in 1917 became the Independent Labour Party}. In the 1908 Federal election, he ran as the S.D.L.P. candidate in the Riding of East Toronto which ran from Parliament Street to Broadview Avenue.

Two years later he ran for a seat in the Ontario Legislative Assembly but he failed to get elected. In both cases, John said that he made a "creditable' showing.

This is John's description of the birth, of the formation of the Social Democratic Labour Party . "On Good Friday 1907 we had a meeting in the Alhambra Hall. We had nine hundred delgates at this Labour Party meeting. Joe Marx of London, a sheet metal worker, sponsored the meeting. Of course, some of us were all fired up - - we were over 900 delegates to that meeting. While there had always been an element of revolt as far as I was concerned against, you know all political parties, there has always been a desire to create a labor party."

"In 1937 you see I ran against Reg Hopkins in Davenport. I ran on a C.C.F. ticket. And a young fellow — Eamon Parks, was my manager. He was a university student, you know It was surprising with the reception I got. It was the first time a straight Labour candidate had run. The Conservative won the by-election."

According to his own account, he didn't make a bad showing. Bruce was the first Labour candidate to run in a Toronto riding. As was his style, John would use every opportunity to spread his gospel. Did so when others lacked the dedication or courage to do so.

"The trades council every time you tried raise political action they talked you off the floor. You could not get anything on the floor about a labour party."

He had more in common with his younger compatriots in the C.I.O. unions, (Congress of Industrial Organizations) most of whom were socialists, unlike his A.F. of L. colleagues, most of whom supported the Tories or Liberals.

John thought highly of Charles H. Millard, National Director of the United Steelworkers of America and former President of the General Motors Oshawa local of the United Auto Workers Both were socialists, non-drinkers and non- smokers.

John Bruce was a pragmatist. He knew the majority of his members were neither temperance advocates, nor pacifists, nor socialists. He was no more successful in convincing his members to support a labor party than have today's labour leaders in getting the majority of their members to vote for the New Democratic Party. He never let his beliefs become divisive, nor did they reduce his effectiveness as a General Organizer.

He saw the old-line parties, Liberals and Conservatives, in Canada, as the hired guns of big business and the establishment. He saw the Democratic and Republican parties in the United States as serving the same masters. He accepted the reality that his members were at the mercy of government, regardless of the party in power.

John knew that to be effective he must be informed; he must be aware of the machinations of all the political parties and of big business. Even as a boy, he was an avid reader. He formed this habit of reading newspapers and magazines when his father had him read aloud to the men that frequented David Bruce's shoe shop in Port Melbourne.

He did not hesitate to comment on the tactics of Canadian Prime

Ministers or U.S. Presidents. He kept a watchful eye on the Members of Canada's House of Commons, the House of Representatives, the Senate and the White House.

Although an avowed pacifist John served his adopted country well in both the First and Second World Wars. For his service to his country in the Second World War George the VI awarded John the title "Officer of the Order of the British Empire." He was the only Canadian unionist to ever receive this award.

Despite his beliefs, John Bruce met with and worked with the establishment when they shared a common interest, or when he was trying to persuade governments to pass laws beneficial to the labour movement, or to the general public.

John Bruce was an eternal optimist. All his life he viewed the world through 'rose-coloured glasses'. He was ready with a program to put what ever went wrong to right again. If he could avoid it he spoke ill of no man, except those whom he thought were a threat to his beloved United Association and the trade union movement.

Before writing his report for the month of December, 1961, John fell down the subway steps in a Toronto subway station and broke his pelvis. At that time, he had every intention of returning to work. After being off work a year, John took his official retirement as of January 1st, 1963.

John's longevity earned him a unique place in Canada's trade union history. During the decade before he retired as General Organizer, he was honoured at conventions, meetings and banquets.

John always kept his family separate and a part. Alice Ripley, John's first wife died in 1912, a year after giving birth to John, the youngest of the four Bruce children. John Bruce remarried in 1913 to the lady who had nursed Alice in the last year of her life. His second wife was a widow with no children. John described her as having been an attractive vivacious French Canadian lady. Yet in the thirty some hours of his taped interviews he never mentioned her name. She died in 1942 of a heart attack.

John had three daughters. Alice and Hazel who were born in Australia married and lived in and around Toronto all their lives. The youngest daughter Amy born in South Africa and never married. She lived with John until he died. She said her father scared away the young men who came to court her. John, the son born in Toronto, left the city as a young man and moved to Connecticut. He was an artist who became an industrial designer.

www.ingramcontent.com/pod-product-compliance
Lightning Source LLC
Chambersburg PA
CBHW022056210326
41519CB00054B/494

* 9 7 8 1 5 5 2 1 2 4 5 7 4 *